About Island Press

Since 1984, the nonprofit organization Island Press has been stimulating, shaping, and communicating ideas that are essential for solving environmental problems worldwide. With more than 1,000 titles in print and some 30 new releases each year, we are the nation's leading publisher on environmental issues. We identify innovative thinkers and emerging trends in the environmental field. We work with world-renowned experts and authors to develop cross-disciplinary solutions to environmental challenges.

Island Press designs and executes educational campaigns in conjunction with our authors to communicate their critical messages in print, in person, and online using the latest technologies, innovative programs, and the media. Our goal is to reach targeted audiences—scientists, policymakers, environmental advocates, urban planners, the media, and concerned citizens—with information that can be used to create the framework for long-term ecological health and human well-being.

Island Press gratefully acknowledges major support from The Bobolink Foundation, Caldera Foundation, The Curtis and Edith Munson Foundation, The Forrest C. and Frances H. Lattner Foundation, The JPB Foundation, The Kresge Foundation, The Summit Charitable Foundation, Inc., and many other generous organizations and individuals.

Generous support for the publication of this book was provided by Decker Anstrom and Sherron Hiemstra.

The opinions expressed in this book are those of the author(s) and do not necessarily reflect the views of our supporters.

Lyme

Lyme

THE FIRST EPIDEMIC OF CLIMATE CHANGE

Mary Beth Pfeiffer

ISLANDPRESS

Washington | Covelo | London

ISLAND PRESS is a trademark of the Center for Resource Economics.

Library of Congress Control Number: 2017958888

All Island Press books are printed on environmentally responsible materials.

Manufactured in the United States of America
10 9 8 7 6 5 4 3 2 1

Keywords: Borrelia burgdorferi, bull's-eye rash, babesiosis, chronic Lyme, climate change, *Ixodes* ticks, Centers for Disease Control and Prevention (CDC), *Bartonella,* vector-borne disease, Infectious Diseases Society of America (IDSA)

To my mom
Kind, strong, joyful

Contents

Acknowledgments

This book was a journey led by science and aided by scientists. I thank them, some named in these pages, many not, for sharing and explaining their work. I thank them for their commitment. Science that is supported, challenged, and considered will ultimately unravel the mystery of Lyme and its related diseases.

I thank the physicians, in North America, Europe, and Australia, who gave me their time, made connections for me, and explained their diagnostic and treatment quandaries. Thanks also to the advocacy groups, in towns, states, provinces, and nations, which are trying to redefine this disease and have supported my efforts. A great many people in this large and growing community have helped me understand what the bite of an infected tick meant for them, their children, and their families. They include Julia's father, Enrico Bruzzese, and Niki's and Keara's mother, Kaethe Mitchell. They include Lyme patients, and their parents, in many states and several countries. I thank them for their time, their stories told in long missives or conversation, and their willingness to provide ever more details and documents. Sometimes, as they and more than a few scientists and physicians will see, all that effort amounted to a single sentence or a pregnant paragraph. But it helped tell the story.

Closer to home, thanks to my husband, Rob Miraldi, who set many things aside to support me in the intensive research and writing process. He made dinners, did laundry, paid bills, but, most of all, listened with the ear of a journalism professor whose mentorship and guidance has long made me a better reporter. Thank you to my children, Sara and Robert, and their spouses, Quinn and Kelsey. They cheered me on at every stage. And, of course, there is my mother, Helen, still going strong and living across the lane from me. I cherish her company and her indignation, which on this subject is considerable.

Thanks to my friends. Janet Graham Gottlieb helped me figure things out in the telling of this story. Raedel Silverman, Kathleen Norton McNulty, and Carolyn Hansen asked good questions. Thanks to my two Dutch sisters, Marianne and Susanne Peters, for translation help and for putting me up, not once but twice, during my research. Thanks to Bob Silverman and Ronnie Liadis for accommodating my research diversions in Cyprus and elsewhere; to Bob and Lisa Brayman for their kindness and encouragement.

I have been blessed with editors who have supported my investigative reporting for the last twenty-five years. Stuart Shinske edited and encouraged the series of articles, published from 2012 to 2015, that led to this book. He and *Poughkeepsie Journal* publisher Barry Rothfeld gave me the time and support that investigative reporting demands. Thanks, too, to Mimi McAndrew, who early on charted my path. I would not be a journalist but for Mimi.

Among many others who helped, I thank Jill and Ira Auerbach, Lorraine Johnson, Dorothy Leland, Pat Smith, Barbara Buchman, Holly Ahern, Jane Marke, Kris Newby, Kenneth Liegner, Betty Maloney, Rosalie Greenberg, Phyllis Freeman, Chris Fisk, Lonnie Marcum, Fred Verdult, Diana Uitdenbogerd, Katherina Deutsch, Eoin Healy, Sandra

Pearson, Michael Cook, Elliot Cowton, Caroline Fife, Rona Cherry, Dana Parish, Andre Efftink, Joy and Kim Collins, and others too numerous to name. Government agencies and their various staff members helped too: The National Institute of Public Health and Environment in the Netherlands, Public Health England, the Public Health Agency of Canada, the US Centers for Disease Control and Prevention, and the US Geological Survey. Thanks to Cary Institute of Ecosystem Studies and Wageningen University in the Netherlands.

To my assistant, Sara McBride: Bless you. To my editor, Emily Turner, thank you for believing in this project and for your direction. To my agent, Rob Wilson, for all your efforts on my behalf, this time and for my previous book, thank you.

Finally, I am grateful to the Internet. How strange is that? It made the world a smaller place, one in which I could, in a single day, speak with people on three continents, access a couple of dozen scientific papers, and check facts that would've taken far longer to look up not too long ago. Through digital communication, I made friends in other countries. I learned that they are like the people I know here in the United States. They are searching for answers. For all they gave me—of themselves and their pain—I hope to show my sincere gratitude by helping to provide some.

—Mary Beth Pfeiffer, Stone Ridge, New York

Introduction

On the day after Christmas in 2015, I took a walk with my thirty-year-old son in an old cornfield that long ago morphed, with changing patterns of agriculture, into a gently tended meadow we know well. This nine-acre patch of earth, across the dead-end lane from our home in upstate New York, has a rare and wonderful feature that we have worked, with the cooperation of neighbors, to sustain: a mowed trail around its perimeter that allows access even when summer mustard, milkweed, and goldenrod are four feet high and the blackberry brambles profuse.

As we often do, we took with us that day a coterie of mismatched family dogs—a Shih tzu, Chihuahua, springer spaniel, and a beige rescue that we call a boxer.

The day was unusually balmy in the last week of a year that had gone down as the warmest in 135 years of weather history, followed only—but considerably—by the year before. The temperature had topped out at 55 degrees Fahrenheit that December 26th; it had reached into the mid-60s on Christmas and hit 72 in nearby Poughkeepsie the day before. For that time of year, daytime temperatures at or near freezing would have been far more typical here in the Hudson Valley, a hundred miles north of New York City. Instead, it had been so warm that forsythia buds had sprouted in one neighbor's garden; crocuses peaked through in another's.

When we came back from our short walk, we did what has become in New York State a routine practice from spring through fall, but not for winter. We checked the dogs for ticks. When we were done, we had picked twenty-one blacklegged ticks from the scruffs of our pups, each about the size of a small freckle, and all with one goal in mind: to latch onto a warm body and suck its blood.

When I moved to this rural county in upstate New York thirty years ago, such things did not happen. Ticks certainly did not show up in December, were indeed rare, and, if seen, were usually of the easier-to-spot dog tick variety, which infrequently carried Rocky Mountain spotted fever. They did not pack the potential arsenal of infection of the small, ubiquitous blacklegged ticks of today. Every year, the list of diseases found within this tick grows longer, including new bacterial, viral, and parasitic pathogens.

These infections have changed daily life in the county in which I live, and they have altered the course of many lives, particularly when they go unrecognized for weeks or months. Mention Lyme disease at a gathering here and nearly everyone has a story. The odd rash, fever, occasional palsy, meningitis, and joint pain of early stages. The sometimes-crippling arthritis, memory loss, depression, numbness, and fatigue of advanced disease.

Even the rare infection that can kill. In towns near me, Lyme carditis, in which the bacterium quickly cripples the heart, claimed a seventeen-year-old high school boy and a thirty-eight-year-old father of three within five months. A woman, who at ninety-one was an active gardener, died after being bitten by a tick and contracting another common tick-borne disease, babesiosis.

Blacklegged ticks have taken up residence in half of continental America's counties, spreading west, north, and south from the Connecticut town for which Lyme disease was named in the late 1970s, like some unchecked algal bloom. These eight-legged arachnids have turned childhood from a time to explore nature to a time to fear it. My seven-year-old grandson has been warned since he could walk: Don't brush up against the tall grasses at the edge of the trail. Don't venture into the weeds. Tuck socks into pants. Sit still for repellent. Check yourself, and remind Mommy to also.

Then, pray we see the tick, and pity the parent who doesn't know to look. Guidance counselors and teachers have told me of children, the most frequently infected, missing months, sometimes years, of school because the tick went unseen or the symptoms were misconstrued. But happenstance is not the only reason that cases are missed or treatment delayed, I have learned. Many patients have suffered needlessly, in the United States, Canada, England, France, Germany, and many other countries, because of how Lyme disease has been framed in American medical journals and treatment guidelines. This is an illness that has been minimized, underestimated, and politicized to the point that doctors fear treating it aggressively with a cheap, common drug: antibiotics.

I began to write about Lyme disease as an investigative reporter for a Dutchess County, New York, newspaper in 2012. I intended to write one or two stories about a prevalent local disease, focusing on its origins, growth, and management by health officials. But Lyme disease proved to be a story far beyond the parameters I'd envisioned. Over a period of four years, I studied the policies, paper trail, and scientific literature. I tested the postulates of patients and their advocates. Many of their assertions, I concluded, were true.

Decades-old testing protocols failed to diagnose many people with the disease. The risk of overtesting—and falsely diagnosing people with Lyme disease—was exaggerated at the expense of cases missed and symptoms dismissed. Official figures did not reflect the epidemic's scope and were soon revised tenfold. Human trials concluded that short-term antibiotics eradicated the bug, while animal and test-tube studies showed they didn't.

And then there was the politics of Lyme disease. Doctors who prescribed repeated courses of antibiotics—standard in other diseases—faced professional disciplinary charges, huge legal fees, and sometimes suspension or loss of their licenses. Research was discredited, ignored, or relegated to second-tier journals when it challenged prevailing dogma. Scientists who were invested in a benign view of Lyme disease used access to elite journals to uphold the status quo. And little money was available for treatment trials because the medical system purported to have the answers to Lyme disease care.

And yet, some 10 to 20 percent of people infected annually with Lyme disease, estimated at 380,000 Americans in 2015, have symptoms that linger months and sometimes years after treatment. The United States Centers for Disease Control and Prevention (CDC) calls the condition Post-Treatment Lyme Disease Syndrome, and it affects multitudes of people as disease-toting ticks move around the world. Lyme disease is rife in most every country in Western Europe. It is in Russia, China, former Soviet satellite states, and though officially unrecognized, in Australia too. In one small study, half the ticks in two parks in New York City harbored the Lyme disease pathogen. In a Chicago park, *Ixodes* ticks, some laden with the Lyme pathogen, outnumbered every other tick carried by migrating birds in 2010; five years earlier, there were none. In a northwestern suburb of Beijing, one in twenty residents tested positive in a Chinese study that said, quite remarkably, "Underdiagnosis of early Lyme disease and physical damage at advanced stage are huge problem [*sic*] in this area."

But infected ticks haven't just moved around. They have added layers of woe. Forty-five percent of ticks in the French Ardennes, for example, carry more than one disease; some have five. A tick-borne malaria-like illness, babesiosis, unheard of not so long ago, became a nationally reportable disease in 2011 after cases skyrocketed in some American states—and it slipped into the US blood supply, infecting premature babies and killing at least eight people. Another rapidly emerging pathogen, *Borrelia miyamotoi*, prevalent in the San Francisco Bay area, has upped the ante on Lyme. Tick babies, hatching by the thousands, can inherit the bug directly from mom rather than get it with its first animal blood meal. Add to this Powassan virus, which is also passed to baby ticks, transmitted within fifteen minutes of a bite, and is fatal 10 percent of the time, and anaplasmosis, a bacterial infection that is particularly dangerous for the elderly. Now consider how a tick infected with two or three of these organisms, for which a doctor may not know to look, and for which testing is not routine, can wreak havoc in a human body.

This spreading toll of tick-borne disease is not a random act of nature. By virtue of the developed world's lifestyle and influence—the cars, the suburban tracts and carved-up forests, the diminished biodiversity, the emissions and airplanes—we have paved the way for the first global epidemic abetted by climate change. Warming may not have caused this scourge, but it most certainly is enabling it.

The pathogen that causes Lyme disease, *Borrelia burgdorferi*, has existed in the environment for millennia, just as blacklegged ticks have long been transported on the wings of birds to new and far-flung climes. What is different is that they now survive there. We have revived and empowered a sleeping giant, the *Borrelia* bug, by helping to create conditions favorable to the ticks that deliver it.

Climate change alone hasn't done this—ticks thrive amid the mice and deer so prevalent in an attenuated, postindustrial era—but its influence in supporting ticks, the vector of human transmission, is key. Ticks have climbed latitudes like ladders; they have moved up mountains. Driven by warmer winters and earlier springs, they are predicted to move well north into Canada in coming years, just as they migrated up the Atlantic seaboard and north into Scandinavia at the close of the twentieth century.

Lyme disease, of course, is not the only epidemic to blossom in our changing world. Illnesses carried by mosquitoes—Zika, West Nile, Chikungunya, dengue fever, malaria—are proliferating and are predicted to worsen as the globe warms. The diseases they carry today annually cause millions of illnesses in tropical countries—and in the case of malaria, nearly a half-million deaths.

When these illnesses arrive on US shores, they capture the public imagination and lead the nightly news in a way that Lyme disease never has. Yet the Lyme toll in the United States and many developed countries is far higher—perhaps four or five million Americans infected to date—than West Nile, Zika, and all other mosquito-borne illnesses combined. And like West Nile, Lyme sometimes kills, although the numbers are largely unknown. Late-stage Lyme disease has even led to suicide, an outgrowth both of the illness and society's regard for it.

For official purposes, Lyme disease is not called an epidemic. It is an "endemic," a term with far less urgency, reserved for an illness that comes to stay. It emerges. It takes root. And it does not leave. It is a slow burn, this scourge, established, entrenched, and for many, inescapable. But its second-class status also stems from how Lyme disease has been managed and shaped. Treatment guidelines issued by the Infectious Diseases Society of America (IDSA) in 2006 diminish the lingering symptoms of Lyme disease sufferers, calling them the "aches and pains of daily living." This, while studies from Columbia University, Johns Hopkins, and elsewhere have measured significant neurological, cognitive, and physical impairments in treated patients. One study equated the quality of life of those with longstanding Lyme disease to people with congestive heart failure. Others have linked the disease to mental illness and showed brains deprived of blood flow.

Borrelia burgdorferi is a clever, adaptable bug. It has a hugely complex genetic profile, with more independently replicating structures, called *plasmids*, than any other bacterium. It doesn't need iron, unlike most other bacteria, removing one option for our immune systems to starve it into submission. It grows slowly, confounding drugs that work on rapidly dividing cells. It hides in places that diagnostic tests do not reach. This tiny spiral-shaped organism is actually a boon to ticks. Those infected are more likely to find a blood meal, and ominously, may even cope better in dryer, hotter conditions, than those that do not carry the Lyme pathogen.

In 2014, a report by the Intergovernmental Panel on Climate Change, a multinational effort, said the period from 1983 to 2012 was likely the warmest 30 years in the Northern Hemisphere of the previous 1,400 years. The report, relying on multiple, independent temperature and climate indicators around the globe, called this "robust multi-decadal warming." Tellingly, the US government monitors Lyme disease as a key indicator of the effect and pace of climate change. Like millions of other people, I see this change in my everyday life. It means that on a short walk in late December, I found twenty-one ticks on four panting, happy dogs that like nothing better than a romp through our shared preserve.

When I look at that lovely field, I see something else. I see a beautiful, inviting menace, a dark and dangerous wood. I see the first epidemic in the era of climate change, long in the making, global, and here to stay. Call it, if you will, the first pandemic. There are things we can do to protect ourselves, to control the bug, to limit its spread. You will read about that in this book. But you will read also about the missed opportunities, the misconceptions, and the human contribution to an epidemic that, for now at least, is beyond our ability to stop.

CHAPTER 1:

Ticks, Rising

Evolution has endowed the big-footed snowshoe hare with a particularly nifty skill. Over a period of about ten weeks, as autumn days shorten in the high peaks and boreal forests, the nimble, nocturnal hare transforms itself. Where it was once a tawny brown to match the pine needles and twigs amid which it forages, the hare turns silvery white, just in time for the falling of winter snow. This transformation is no inconsequential feat. *Lepus americanus*, as it is formally known, is able to jump ten feet and run at a speed of twenty-seven miles per hour, propelled by power-ful hind legs and a fierce instinct to live. But it nonetheless ends up, 86 percent of the time by one study, as a meal for a lynx, red fox, coyote, or even a goshawk or great horned owl. The change of coat is a way to remain invisible, to hide in the brush or fly over the snow unseen, long enough at least to keep the species going.

Snowshoe hares are widely spread throughout the colder, higher reaches of North America—in the wilderness of western Montana, on the coniferous slopes of Alaska, and in the forbidding reaches of the Canadian Yukon. The Yukon is part of the Beringia, an ancient swath of territory that linked Siberia and North America by a land bridge that, with the passing of the last Ice Age 11,000 years ago, gave way to the Bering Strait. All manner of mammals, plants, and insects ferried east and west across that bridge, creating, over thousands of years, the rich boreal forest. But in this place, north of the 60th parallel, the axiom of life colored by stinging cold, early snow, and concrete ribbons of ice has been upended in the cosmic blink of an eye. The average temperature has increased by 2 degrees Celsius in the last half century, and by 4 degrees in the winter. Glaciers are rapidly receding, releasing ancient torrents of water into Kluane Lake, a 150-square-mile reflecting pool that has been called a crown jewel of the Yukon. Lightning storms, ice jams, forest fires, rain—these things are suddenly more common. Permafrost is disappearing.

Such rapid-fire changes across a broad swath of northern latitudes are testing the adaptive abilities of the snowshoe hare, however swift and nimble it may be. Snow arrives later. Snow melts earlier. But the hare changes its coat according to a long-set schedule, which is to say the snowshoe is sometimes snowy white when its element is still robustly brown. And that makes it an easier target for prey. In 2016, wildlife biologists who tracked the hares in a rugged wilderness in Montana gave this phenomenon a name: "climate change-induced camouflage mismatch." The hares molted as they always had. It's just that the snow didn't come. Survival rates dropped by 7 percent as predation increased. In order to outwit its newest enemy—warmer winters—snowshoe hares would need something on the order of a natural miracle, what the biologists, writing in the journal *Ecology Letters*, called an "evolutionary rescue." Like the Yukon, this pristine corner of Montana was projected to lose yet more snow cover; there would be perhaps an additional month of bare forest floor by the middle of this century, on which snowshoe hares would stand out like bright white balloons.

In the tally of species that will evolve or perish as temperatures rise, consider now the moose. The lumbering king of the deer family, known for antlers that can span six feet like giant outstretched fingers, faces a litany of survival threats, from wolves and bears to brain worms and liver fluke parasites. But in the late 1990s in many northern states and Canada, something else began to claim adult cows and bull moose and, in even greater numbers, their single or twin calves.

Lee Kantar is the moose biologist for the state of Maine, which means he makes a living climbing the rugged terrain of north-central Maine when a GPS collar indicates a moose has died. A lean man with a prominent salt-and-pepper mustache who wears flannel shirts and jeans to work, Kantar tagged sixty moose in January of 2014 around Moosehead Lake in the Maine Highlands. By the end of that year, twelve adults and twenty-two calves were dead—57 percent of the group. When biologists examined the carcasses, they found what they thought was the cause. Calves not even a year old harbored up to 60,000 blood-sucking arthropods known as winter ticks. In Vermont, dead moose were turning up with 100,000 ticks—each. In New Hampshire, the moose population had dropped from 7,500 to 4,500 from the 1990s to 2014, the emaciated bodies of cows, bulls and calves bearing similar infestations of ticks. These magnificent animals were literally being bled to death.

Winter ticks have been known to afflict moose since the late 1800s. In a normal year, a single moose might carry 1,000 or even 20,000 ticks. In a particularly harsh winter, when moose are underfed and weak, anemia and hypothermia wrought by ticks can make the difference between life and death. Bill Samuel, a retired University of Alberta biology professor, has spent a career studying the moose of North America. He painstakingly counted 149,916 ticks on a moose in Alberta in 1988. In a 2004 book, he recounts episodes of ticks killing moose in Saskatchewan in the spring of 1916, in Nova Scotia and New Brunswick in the 1930s, and in Elk Island National Park in central Alberta at points from the 1940s through the 1990s. Some of the animals were so infested that there was not a tick-free spot in the arachnids' favored places—the anus, the inguinal area, the sternum, the withers and lower shoulders. In futile attempts to rid the parasite, these pathetic animals had rubbed against trees to seek relief, losing long, lustrous fur and leaving grayish, mottled patches. They are called "ghost moose."

Moose have long died from disease, predators, hunting, and sometimes ticks. But their losses in the early twenty-first century had a different, more threatening, more consequential implication. In 2015, two environmental organizations, alarmed at population trends, petitioned the US Secretary of the Interior to have the Midwestern moose listed as an endangered species. In Minnesota, the number of moose dropped 58 percent in the decade through 2015, similar to losses in New England. Environmentalists believe moose could well be eradicated in the Midwest by 2020, with stocks declining precipitously in Wisconsin, Minnesota, and Michigan.

Lee Kantar knew that ticks were killing his moose in Maine. What's becoming clear is why winter ticks had infested his herd, draining half their blood from every available patch of skin. "The greatest threat confronting the species," declared the Center for Biological Diversity and Honor the Earth in the 2015 petition to help moose, "is climate change." Not hunters. Not habitat loss. Not even pollution, though that is important. Moose like and need the cold. They become sluggish when it's warm, failing to forage as they should and becoming weak and vulnerable. In the warmer, shorter winters of the US Midwest and Northeast, bumper crops of winter ticks are surviving to wake up when the trees burst to life in earlier springs; they have more time in longer falls to cling in veritable swarms on the edges of high bushes, their legs outstretched, waiting for a ranging, unsuspecting, and wholly unprepared moose. When the moose lay in the snow, they leave carpets of blood from engorged ticks. When a baby moose emerges from the womb in Minnesota, a band of thirsty ticks moves from mother to neonate. The moose shed those fat, flush ticks onto fall and winter ground, and the ticks snuggle into the leaf litter rather than freeze in the snow, as they once might have, reducing tick mortality but upping that of the moose.

Bill Samuel is a careful scientist who does not jump to conclusions, and he sees many forces working together to kill off moose in the finely tuned orchestra that is the outdoors. Wolves, liver fluke, brain worms, unmanaged hunting, habitat loss—they are all part of the picture. Because of how it affects and is affected by those other factors, "climate change," he told me, "might be the major one."

"It's the Ticks"

Jill Auerbach knows that the winter ticks attached to dead and dying moose pose little threat as a species to humans, whom they aren't prone to bite. But when news broke of moose losing half their blood to winter ticks, she was horrified and worried. Auerbach, an active woman in her seventies, was bitten in her forties by a small tick that thrives in the woods, thickets, and backyard edges of the county in which she lives, in New York State's Hudson Valley. She lost ten years of her life to that tick, had to retire as a highly rated programmer at the nearby IBM plant, and still suffers the aftermath of a case of Lyme disease that was caught too late. "It brought me to my knees," said Auerbach, among an all-too significant share of people infected with Lyme who suffer long-term symptoms. To her, the rise of winter ticks is one more indicator of an environment out of whack, and so is the more measured, but nonetheless relentless, surge in blacklegged ticks, like the one whose bite haunts her thirty years on.

That other tick, known to scientists as part of the *Ixodes* genus—in Auerbach's case, *Ixodes scapularis*, or blacklegged tick—is spreading across the United States and in many other countries with startling alacrity. Canada, the United Kingdom, Germany, Scandinavia, Inner Mongolia in China and the Tula and Moscow regions in Russia: they are all grappling with large and growing numbers of disease-ridden ticks. Infected ticks have been found in urban parks in London, Chicago, and Washington, DC, and in the open, green expanses of Killarney National Park in Ireland's southwest. In Western Europe, where case reporting is not standardized, the official case count is 85,000 per year; a 2016 analysis, published in the *Journal of Public Health* at Oxford, England, put the number at 232,000. Signs of a burgeoning problem are apparent in Japan, Turkey, and South Korea, where Lyme cases grew from none in 2010 to 2,000 in 2016. When I asked three Spanish physicians in 2017 where Lyme disease was found in Spain, one said, "everywhere," and the others agreed. One of them, Abel Saldarreaga Marín, had treated forestry workers in Andalucía, where he said symptoms are often managed, perilously, with traditional remedies. In the Netherlands, as elsewhere, warnings to protect Dutch hikers, children, and gardeners from bites had failed for years to curb the growing toll, then hit what may simply have been a saturation point, with *Ixodes ricinus* ticks inhabiting 54 percent of Holland's land.

Across the Atlantic Ocean from Holland, the US Centers for Disease Control and Prevention (CDC) in Atlanta issues maps every year showing, by virtue of small black dots, the presence of Lyme disease cases in American counties. The CDC's 1996 map was the first to officially chart US Lyme cases, although the disease was well along by then. Dots on that inaugural map collectively create an unremitting black smudge along the Atlantic shore from Delaware to Cape Cod. New Jersey, Connecticut, Massachusetts and the lower reaches of New York State—where Auerbach contracted her case—are all inky black. A broken shadow runs along the Wisconsin-Minnesota border, too, with a handful of dots in many heartland states.

But it is the change over the course of eighteen years of maps that is telling, depicting the flowering of Lyme in a sort-of cartoon flip book style as it spreads across the Northeast and Midwest of America. North it goes up New York's Hudson River Valley and into the state's Adirondack Mountains, jumping the border to Vermont's Green and New Hampshire's White Mountains. West and south it moves great guns into Maryland and northern Virginia. By 2014, the dots consume much of Pennsylvania and darken New York's Southern Tier to the shores of the Great Lakes and the St. Lawrence River. The Upper Midwest is liberally peppered. Dots appear in many other states too.

In 1996, blacklegged ticks were known to be established—meaning there were enough counted to breed or they already had—in 396 American counties. By 2015, researchers at the Centers for Disease Control reported the ticks were ensconced in 842 counties, an increase of 113 percent. Remarkably, the study's twin US maps chart the forward march of ticks in much the way that the maps of Lyme cases plot the progression of disease. Both are relentless.

Auerbach, who became a Lyme disease expert and advocate after her long-ago bout, has for years ended her emails with, "What's the problem? Well it's the ticks of course!" They must be stopped, she believes, and the 2015 CDC map shows why. In it, ticks are seen moving into places that only a decade before had been considered ill-suited to support them, from the Allegheny Mountains to the Mississippi Valley, from western Pennsylvania south and east across Kentucky and Tennessee. In Minnesota and Wisconsin, *I. scapularis* "appears to have expanded in all cardinal directions," the CDC researchers reported in language that was sometimes remarkable and alarming. The ticks have "spread inland from the Atlantic seaboard and expanded in both northerly and southerly directions," they wrote, stopped only to the east by the Atlantic Ocean. Tick movement up the Hudson Valley is "recent" and "rapid," the researchers wrote, their expansion overall, "dramatic." Where there had once been a divide between infestations in the Northeast and Midwest, they concluded, ticks merge "to form a single contiguous focus…a shifting landscape of risk for human exposure to medically important ticks."

On the Move

Lyme disease emerged in coastal Connecticut in the 1970s, where symptoms akin to rheumatoid arthritis were reported in a circle of children unfortunate enough to be trailblazers of a disease in which early treatment is key to recovery. Diagnoses made late can portend long and difficult sieges of illness—fatigue, joint pain, learning problems, confusion, and depression. The parents and guidance counselors of Lyme children, and the children themselves as young adults, have told me of school years lost to the disease. Children five to nine years old have the highest per capita Lyme infection rate in the United States, while people sixty- to sixty-four-years old have the highest hospitalization rates for it, according to a study of 150 million US insurance records from 2005 to 2010.

The story of the emergence of Lyme disease now, of its rise in dozens of countries around the world and of millions made sick, must be told through the lens of a modern society living in an altered environment. In the last quarter of the twentieth century, a delicate array of natural forces indisputably tipped—were tipped, more accurately—to transform Lyme disease from an organism that lingered quietly in the environment for millennia to what it is today: the substance of painful stories shared between mothers; a quandary for doctors who lack good diagnostic tests and clear direction; the object of rancor over studies that discount enduring infection while acknowledging persisting pain.

The US Centers for Disease Control and Prevention defines the word "endemic," which it applies to Lyme disease, as the "constant presence and/or usual prevalence of a disease...in a population within a geographic area." But while Lyme is firmly rooted in thousands of locales, it is surely not confined there as climate changes, ticks move, and cases mount. The CDC's designation, even if scientifically accepted, is unfortunate. It serves to minimize the import of a disease that yields some 300,000 to 400,000 new cases in the United States each year, is found in at least thirty countries and likely many more, and is growing precipitously around the world. Lyme disease is moving, breaking out, spreading like an epidemic.

The ticks that carry Lyme disease are, like spiders, arachnids not insects. Although they cannot fly or jump, they are, for all practical purposes, climbing mountains, crossing rivers, and traversing hundreds, even thousands, of miles to set up housekeeping. These feats are documented by scientists who are ingenious at finding ways to track and count ticks. They drag white flannel sheets across leafy forest carpets, sometimes infusing them with piped-in carbon dioxide, the mammal gas that makes ticks reach up, forelegs outstretched, to snag a passing meal. They catch avian migrants infested with hitchhiking arachnids. They count ticks on the ears of trapped mice and shrews, sometimes getting bitten in the process. They dissect bird nests, reach beneath leaf litter, and scour grassy sand dunes.

When these researchers are lucky, they find data from some other era that proves their hunch that something has changed. In 1956, a scientist named Cvjetanovic´ in the Bosnian region of the then Socialist Federal Republic of Yugoslavia reported that *Ixodes ricinus*, the castor bean tick, could not survive at altitudes higher than 800 meters above sea level, or about 2,600 feet. But when Jasmin Omeragic of the University of Sarajevo took another look in 2004, collecting 7,085 castor bean ticks in the Dinaric Alps of Bosnia and Herzegovina, he found them living comfortably at 1,190 meters, or 3,900 feet. In 1957 in Sumava, in then-Czechoslovakia, researchers found the ticks could not survive at elevations above 700 meters. By 2001, biologists found them thriving at 1,100 meters. What those early observations pointed to, wrote Joylon Medlock and his colleagues in 2013, is "clear evidence of an altitudinal expansion of *I. ricinus*." Put another way, ticks are aggressively moving up. But they are also moving in other ways—and to places more suited than steep slopes to human habitation.

In the Hudson Valley of New York State, a team from the University of Pennsylvania used *Ixodes* DNA to draw a family tree of black-legged ticks, much the same way that people use saliva swabs to search for distant ancestors in their genetic code. Studying ticks collected at four locations from 2004 to 2009, the researchers recreated a 125-mile upriver tick migration, similar to that of the colonial Huguenots and Livingstons three centuries earlier. The tree begins in southernmost Yorktown, where the tests showed the ticks residing, give or take, for the previous fifty-seven years. Then, seventeen years later, these eight-legged pioneers climb the next rung north, to bucolic Pleasant Valley. Eleven years pass, and they settle in Greenville, in the foothills of the Catskill Mountains, and, seventeen years later, emerge in northernmost Guilderland, where Dutch settlers from New Netherland had settled in 1639. While other DNA literally crept in along the way—mate-searching ticks do follow their hearts—by far the most dominant strain at each point along the march was the one from southernmost Yorktown. The DNA, the researchers wrote, "strongly support a progressive south-to-north expansion." Defying the odds, the ticks had moved to places where it had long been colder and snowier. And they did just fine.

In Europe, ticks are on a similarly relentless march north. In Sweden, researchers studied the range of the castor bean tick from 1994 to 1996 by dragging cloths in fifty-seven locations and querying residents about bites and sightings. They were able to establish a boundary line at about 60°5'N, above which the ticks could not survive. By 2008, the ticks were found to have moved some 300 miles north, mainly along the Baltic coast, to about 66°N. In Norway, the story was repeated. Twin surveys in 1943 to 1983 found the ticks unable to survive north of 66°N. By 2011, they had traveled 250 miles, to the highest known latitude in Europe, 69°N, Oslo researchers reported, in a record that seems destined to be, if not already, broken.

Nicholas Ogden is senior scientist in the National Microbiology

Laboratory in the Public Health Agency of Canada. He has watched over the last two decades as blacklegged ticks have leaped the US border in a northerly trek, some 600 miles into Canadian territory. In 1990, the only documented location in Canada where the tick was found was in southern Ontario, in a town called Long Point, which is located on a thin strip of land jutting into Lake Erie and much closer to New York State than to Ottawa, Toronto, or Montreal. Less than two decades later, the ticks had established themselves in a dozen more Canadian locations, including in Manitoba, southeastern New Brunswick, and Nova Scotia. In 2008, Ogden and his colleagues mapped the risk of ticks moving north and predicted "possible widespread expansion" into south central Canada. By 2015, another study put the forecast farther: Lyme-toting ticks would move about 150 to 300 miles north by 2050. That puts Canada in much the same position as the United States in the 1980s, and Ogden knows this. The world's second largest country, which saw homegrown Lyme cases grow twelvefold from 2009 to 2013, is facing a burgeoning epidemic of Lyme disease. "It is becoming a real public health problem," he told me.

In 2015, Ogden and his colleagues employed a novel way to track the destination of ticks on migrating birds. Enter the gray-cheeked thrush, a plain, medium-sized bird and determined skulker that hides in the underbrush, making it prone to collect ticks. Ogden's team captured the thrush—along with seventy-two other tick-infested birds—as it crossed the Canadian border on its northward migration. Researchers then studied the molecular composition of its delicate, metal gray tail feathers. These rectrices, which help steer the bird in flight, bear a certain fingerprint, an isotope signature from the hydrogen in the water where the bird fledged. Knowing that birds usually return to the place of their birth, scientists concluded the thrush was destined for the farthest reaches of the study, which covered northern Ontario to the southern Canadian Arctic. Charles Francis, who monitors bird populations for the Canadian Wildlife Service, helped in the study. "Very likely there have always been ticks being introduced to northern areas because of migrating birds," he said. Only now, more of the ticks they carry are surviving in more places. By 2017, Canadian researchers reported that large swaths of Ontario had converted, as a paper in the journal *Remote Sensing* put it, from "unsustainable to sustainable" for Lyme-toting ticks.

While the snowshoe hares struggle in the rugged Montana wilderness, ticks and their pathogens are thriving in a warming world, colonizing more places and multiplying there, just as they did in the last great, post-Ice Age warming. Thirty years ago, health officials in Canada told people with Lyme disease that they had almost certainly acquired the infection elsewhere, usually in travel to the United States. By the first decade of the twenty-first century, they had started to hedge their bets.

A Singular "Indicator"

In 2014, the US Environmental Protection Agency issued a 112-page report on the future of the United States in a warmer world. It began with a conclusion that had been denied, discounted, and politicized in the states for decades, but at last, or perhaps for the moment at least, was accepted as true. "The Earth's climate is changing. Temperatures are rising, snow and rainfall patterns are shifting, and more extreme climate events—like heavy rainstorms and record high temperatures—are already taking place. Scientists are highly confident that many of these observed changes can be linked to the climbing levels of carbon dioxide and other greenhouse gases in our atmosphere, which are caused by human activities."

The report consisted of six sections that attempted to describe and quantify the effects of global climate change—on oceans, on the earth's glaciers, on forests and lakes, and on people. In the report's third edition in 2014, the agency included four new "indicators" to track and measure the impact of climate change. These included the number of annual heating- and cooling-degree days (which are showing Americans using more energy to cool rather than heat); incidence of wildfires; the water level and temperature of the Great Lakes; and last, Lyme disease.

From this point forward, the agency would track the rate of reported Lyme disease cases across the United States as an official outgrowth and barometer of climate change. The tick-borne illness, with perhaps four million American cases since 1990, is the only disease to be accorded that dubious distinction. In discussing direct health impacts of a warmer earth, the agency cites two other trends to watch: heat-related deaths, which were estimated at 80,000 in the last three decades, and ragweed pollen seasons that cause painful allergies for millions. But Lyme disease has a singular distinction. It is an illness spread by ticks, the EPA report states, whose "populations are influenced by many factors, including climate."

In states from Maine to Florida and New York to California, across the breadth of southern Canada and in many parts of Europe, once-sweeping woodlands have been reduced and divided, often into idealized forest fragments at the periphery of residential tracts—places where people can be close to, support, and observe wildlife. Multitudes live, work, and play in or near these green spaces in a new epoch tentatively called the Anthropocene, the era marked by the hand of humanity. The irony is that these adulterated slices of nature and de facto nature preserves are incubators, in many of these places, of Lyme disease. The smaller the patch, in fact, the higher the proportion of diseased ticks, as documented in a study in Dutchess County, New York, where the per capita rate of Lyme disease is among the world's highest.

In these fragments, small mammals, like white-footed mice in North America and garden dormice in Europe, have found havens, thriving in the absence of predators like foxes. In the language of tick-borne disease, the mouse is quaintly called a "host" for ticks and a "reservoir" of Lyme disease, the place where baby ticks, almost too small to be seen, get their first sip of infection. In city parks, suburban tracts, and exurban preserves, people come skin to skin with these ticks. In scores of studies, other environmental factors besides climate change, many controlled by human beings, are pointed to as drivers of this epidemic. The slicing and dicing of forests, and the loss of biodiversity that followed, is surely high on a complex and evolving list.

While there is no single explanation for the twentieth-century emergence of Lyme disease, there is ample evidence that climate change has played no small part. Consider New Hampshire. At Pinkham Notch, a mountain pass along the Appalachian Trail, snowfall has declined an average of four inches every decade since 1970, and days below freezing have dropped by three per decade since 1960. Lilacs bloom earlier in New Hampshire, and the growing season is two to three weeks longer than in 1970. This ruggedly beautiful Northeast state had the United States' second highest rate of Lyme disease in 2013, with a 400 percent increase in cases since 2005. Other research suggests why.

In the Krkonose Mountains in the northern Czech Republic, temperatures increased by 1.4 degrees Celsius in four decades, and *I. ricinus* ticks survive as high as 1,299 meters above sea level. "They didn't decide to go climbing," a scientist there named Michail Kotsyfakis told me. "It's just that they can survive in these areas." In the Montérégie region of southern Quebec, extending south from Montreal to the Saint Lawrence River, temperatures have risen since the 1970s by 0.8 degrees Celsius, and white-footed mice have thrived in shorter, warmer winters. "Its range is rapidly shifting poleward," Canadian researchers wrote in 2013, pointing to "an increasing body of empirical evidence to support the hypothesis that climate warming is a key driver of Lyme disease emergence, acting upon many levels of the transmission cycle of the disease."

The questions are these: Did a changing climate cause this epidemic? Or is climate change merely driving this sickness—with the ticks and animals that circulate it—to new places and new peoples? Evidence most certainly supports the latter. The former is trickier. But Lyme disease is distinctive as the first disease to emerge in North America, Europe, and China in the age of climate change, the first to become entrenched, widespread, and consequential to multitudes of people. It is growing, too, in places like Australia, where residents are told, as they were in southern Canada and still are in many parts of America, Canada, and Europe, that they must have some other illness besides Lyme disease or, if not, they contracted the infection somewhere else. "We're an island. We have island thinking," said a country GP from the mid-north coast of New South Wales named Trevor Cheney, who routinely diagnoses Lyme disease though doctors are told it doesn't exist in Australia. "As if migratory birds"—which drop ticks far and wide—"don't come there," he told me at a conference in Paris.

Such poor advice has cost many Lyme patients valuable time to seek treatment. It stems from a failure, by public health and medical experts, to see the past as future. Lyme disease is moving to new places, as it has for nearly half a century. In the decades since the children of Lyme were infected, little progress has been made to control ticks, protect people from bites, test with certainty for the Lyme pathogen, called *Borrelia burgdorferi*, and especially, adequately treat the infected.

Ixodes ticks, blacklegged, castor bean, or otherwise, deserve our respect. They come armed not only with Lyme disease but with a growing menu of microbes: bacterial, viral, and parasitic, known and yet unnamed. Ticks can, and sometimes do, deliver two, three, or four diseases in one bite. So resourceful are infected ticks that two feeding side by side on the same animal may pass pathogens one to the other and never infect the host. So clever is the Lyme pathogen that infected ticks are more efficient at finding prey than uninfected ticks. These ticks may not be able to fly or jump or trek more than a couple of human steps. But they have changed many lives, cost billions in medical care, and colored a walk in the woods or a child's romp in the grass, our very relationship with nature, with angst.

This is all the more disturbing when we realize, ultimately, that it is we who unleashed them.

CHAPTER 2:

"Invisible Assassin"

She is a young healthy woman in the photo, glossy blond hair curling softly on her shoulders, red lips smiling, blue eyes shining. She looks confident and smart in a black blazer and white blouse, a glimpse of red reference books on a shelf behind her. Barbara Pronk is at her peak here, a time, long ago, when she was happy and looking forward.

Barbara had a good job as a personal assistant at Royal Dutch Shell in The Hague, an accomplishment in a country where professional work, at a multinational company no less, was not easy to come by. She had good friends, colleagues who knew they could count on her, and a mother who adored her. She was funny, ambitious, and beautiful, too, a slim, graceful young woman who dressed for work in much the way that she did her job. Impeccably.

Although she grew up and lived in a small Dutch city called Rijswijk, Barbara loved to travel in the worldly way of many in the Netherlands. There, in one of the earth's most densely populated countries, where existence is defined by a struggle to hold back the sea, children learn English early and are schooled to see themselves as a small part of a big world. It was a world Barbara embraced.

When she was in her twenties, Barbara spent three months living with friends near Clearwater, Florida, where she had planted a garden, a Dutch passion, and gone to the beach. During a sabbatical from Shell, she had lived in California for several months, honing her English in a college course in Santa Barbara and sealing her love for the states. An only child, she took regular cruises with her mother, Josephine van der Ven, who was also her good friend. They were so close that Barbara at one point had bought her own apartment but quickly sold it to move back with Josephine.

In 2005, as her career and life were flourishing, Barbara began to get sick. She suffered overwhelming fatigue and all-over body pain. She developed skin problems and starting losing hair. At the time, Barbara assumed it was the stress of a demanding job as secretary to the comptroller in Shell's oil and gas unit. When she fainted while shopping and needed first-aid, she again wrote it off to overwork. She was too busy, she had too much to achieve, to be ill. The illness had other plans, however.

Over a period of six years, Barbara and her mother would travel widely and spend thousands of euros in her search for a diagnosis and a cure. Her symptoms, like those of many such patients, were what medicine calls "nonspecific"—not characteristic of a particular disease but associated with a long list of them. She suffered unremitting joint and muscle pain. She was hypersensitive to light, touch, stress, and sound. Once acutely attuned to detail, she could not remember things but was keen enough to know what it suggested: dementia. In her search for help, Barbara went to some thirty doctors and clinics in the Netherlands and other countries.

Some doctors said she had myalgic encephalomyelitis or chronic fatigue syndrome, labels that more aptly described symptoms rather than disease. Others doctors were less kind, suggesting the problem might be, as Josephine put it, "between her ears," or, as Barbara told a friend, "attention seeking." Barbara did not see it that way, and she started doing research of her own. It made her remember something. There was the time in Florida, a few days after working in the garden. She had seen a radiating red rash on her leg—she had even emailed her mother about it. At the time, Barbara thought of seeking out a doctor, but it would have been difficult in a foreign country. She dismissed the idea, and the rash went away. Years later, sitting at a computer, trolling for diseases with symptoms that matched her own, Barbara thought she had found her diagnosis in the way that many advanced Lyme disease patients do.

Deceptive Tableau

The tidy, affluent country that is the Netherlands is as crowded as, though more livable than, countries like Bangladesh, Rwanda, and India. Its 17 million people live toe-to-toe in well-kept villages, planned suburbs, and cities buzzing with bicycles and commerce. The landscape is flat, low, often reclaimed from the sea, and well used—an amalgam of farm fields, polders, canals, grasslands flecked with sheep, and precious fragments of second-growth forest. The Dutch mourn their environmental plight. They spout statistics on crowding, refer longingly to the wide-open spaces of America, and debate the merits of recycling and modern wind turbines.

So hungry are Holland's people to commune with the land that railroad edges are lined with small fenced plots on which rows of lettuce, beans, and flowers are nurtured, and tiny sheds are dressed up as mini-dachas in the country. The Dutch pitch tents on sandy seaside hills and in Friesland woods by the score. They bicycle along the Rhine Canal and down brushy rural paths. They use any patch of warmth and sun, in a rainy country, to dine and sip coffee outdoors.

The Dutch recipe for preserving landscape and life-quality has made the Netherlands a lovely, habitable land—I lived there in 1991 and visit often—and an incubator for Lyme and tick-borne disease. Forget the tulips and windmills of tourist pamphlets. Here is where ticks toting dangerous pathogens mingle with and infect human beings at rates that, in some towns, are as high as anywhere in the world. On the windswept island of Terschelling, located in the Wadden Sea off the north coast, five of every thousand residents are infected yearly, rivaling disease prevalence in the United States' Northeast, where Lyme was first recognized in the 1970s.

The ingredients for this Dutch epidemic of tick-borne disease are cloaked in a charmingly deceptive natural tableau. Suburbs abut purple fields of heather inviting walkers onto paths lined with crushed seashells. Thrushes and small mammals thrive in a regulated, postindustrial environment. Ruddy roe deer, perhaps 90,000 nationwide, prance in fields and jump hedgerows, so common they collide with motor vehicles some 10,000 times a year.

Willem Takken, a white-haired, white-mustachioed entomologist at Wageningen University, in 2001 began studying the dunes along the Netherlands' blustery North Sea coast. He found them "riddled with ticks," as he put it, with one in ten harboring the Lyme disease pathogen. He realized then the need for a national project to track the movement, infection rates, and profusion of *Ixodes ricinus*, the species of European tick that delivers the Lyme punch. Starting in 2006, he and his university colleagues began an annual tick census at fifteen sites across the country. Their research tells a tale of infestation and disease, with growing numbers of ticks infected with Lyme disease with each passing year.

In 1865, a novel about Hans Brinker and his quest to win a pair of silver skates—and make his injured father well—made Holland's frozen canals legendary. But those icy highways are a distant memory. The nation's historic eleven-city skating race—a 120-mile trek as storied as the Alaskan Iditarod—had not been held as of the winter of 2016–17 in twenty years. It simply had been too warm. In the last century, temperatures have gone up 1.7 degrees Celsius, or about 3 degrees Fahrenheit, in the Netherlands and summer-like days have increased by nearly twenty. The potential for rising temperatures—and oceans—isn't lost on a country with a finely tuned system of dykes and pumps, where 17 percent of the land has been reclaimed from the sea.

In a warmer Netherlands, there is virtually no time of year when it is consistently cold enough to guarantee against questing ticks, namely those in search of one of the blood meals *Ixodes* ticks take at each of its three life stages. Indeed, instead of taking two or sometimes three years to complete their life cycle, ticks now may live out their lives in half the time, skipping or shortening those dormant times known as diapause. Up to 10 percent of the Dutch population—or 1.7 million people—likely would test positive for antibodies that show exposure at some time to the Lyme disease pathogen.

The irony is that this onslaught of diseased ticks is the product of two colliding forces: the manufactured, pollution-driven phenomenon we call climate change, and human efforts to revive and coexist with a natural world we have altered. When he was a boy growing up in Holland, WillemTakken rarely saw a bird of prey. Today, they soar over a landscape with wild boar and deer, reassuring signs to the Dutch that nature survives amid industrialization, after DDT, and despite the incalculable violation of World War II. "We now see a wildlife population that is actually as it was in the nineteenth century," he said.

But this modern version of nature is different, sanitized, unbalanced. In today's Holland—and in many other developed lands—people live in what a Yale University researcher wryly named the "woodburbs," happily cohabitating with and admiring the wildlife around them. But the animals of this contrived green landscape, most especially mice and deer, come with a catch. They are essential to the survival and sustenance of ticks, and there are few natural enemies, as there once might have been, to keep the animals in check. The European roe deer bordered on erad-ication in the early twentieth century. But protected by hunting limits and the love of an adoring public, its population surged by 50 percent in Europe in twenty years through the early 2000s, to about nine million animals.

Ticks generally pick up the Lyme infection, in their first and second stages of life, mostly from small mammals, in particular mice (including, in Holland, the garden dormouse). These pesky rodents once might've been eaten by hungry foxes, animals that in this scenario at least are few and far between. When these diseased ticks molt into adults, mice and other small mammals will not do. It is on the flanks or ears or eyelids of deer that ticks find their purpose in life. They feed, they breed, they fall off, and the cycle ends with each pregnant female laying several thou-sand eggs to support a new *Ixodes* generation.

The birds move ticks around. Warming climate helps them survive. Mice infect them. But, said Takken, "If there were no deer," to sup-port adult ticks, "there would be no Lyme disease." That is debated, as is most everything in a disease in which truth and fact, as we will see, are evolving.

"Shitty Disease"

The road to diagnosis was long and difficult even after Barbara Pronk thought she knew what was wrong. There were many, many tests, "cheap and worthless," her mother Josephine said, before a laboratory in Germany, using an alternative to the standard assay, confirmed Bar-bara's suspicion: she had Lyme disease, and, because of the diagnostic delay, it was of the worst kind. Long-established, resistant to treatment, advanced. In such cases, the Lyme bacterium, a spirochete known as *Borrelia burgdorferi*, finds ways to evade the body's immune system and survive an onslaught of medications. It damages tissues, causing last-ing pain. It hides dormant, in places diagnostic blood tests don't reach, giving its host a sense of wellness; then it strikes again. Barbara called the disease an "invisible assassin," a description with which many late-stage Lyme patients would agree. An upstate New York doctor, speaking to about 400 people at a research forum in Albany, New York, in 2016, put it somewhat less poetically: "It's a shitty disease," said Dr. Ronald L. Stram. In the front row at the forum was a twelve-year-old girl in a pretty dress and a wheelchair, who had been blessed by Pope Francis on an airport viewing line some months earlier. She, like Barbara, had been told that perhaps there was nothing wrong at all even though she had gone to a doctor with the Lyme rash and wasn't treated. Perhaps, it was suggested, she just did not want to walk.

Barbara struggled to capture the dimensions of her illness in an email. "This is not a 'normal illness' but an opponent of unprecedented size: a battle against bacteria and their toxins that plague your body," she wrote in Dutch, "into every organ, tissue, cell, and take you physically and mentally and make you completely devastated inside."

Long after her diagnosis, she shared her illness over coffee with a colleague. The woman put a comforting hand on Barbara's arm. "Please, don't," Barbara said." It hurt to be touched. Like others at Shell, the woman had not known Barbara was ill. When she had previously passed up a promotion to the top tier of Shell secretaries, she told the colleague only that she did not want her job to take over her life.

But something else already had, robbing Barbara of her present and future. Her diagnosis, she wrote, came "much too late." She had pain, "excruciating…unbearable," that no drug could dent. She had had repeated courses of antibiotics, taken herbs and supplements, undergone acupuncture, and tried special diets. Accompanied by her mother, she had traveled to Bali for ozone therapy that, like everything else, did not help her. She cried out regularly in her sleep. She lost hope.

In March 2014, in a flat overlooking the sprawling harbor of

Scheveningen, a resort city on the North Sea not far from her home, Barbara shared champagne with her mother, two of her closest friends, and a psychologist. It had been nine years since she was infected by a diseased tick. They shed tears and held each other. Earlier, they had taken Barbara to the seaside promenade, lined with kitsch and cafes on one side, the city's immensely wide, tawny sand beach on the other. The sun was shining brilliantly, atypical of the Netherlands in March, and Barbara looked out at the sea. The now thirty-nine-year-old woman was a shadow of the one in that sparkling photograph but she was more beautiful than she had ever been, her mother thought, and she was filled with joy. The decision had been made.

Barbara had drafted an email to twenty-six members of the Dutch Parliament's House of Commons, along with prominent Lyme disease doctors and researchers, Shell executives and coworkers, journalists, and others. It was Barbara's attempt to make what would happen to her, and had already happened to her, matter. She would succeed, with the ironic luck of good timing and the work of many people who, like her, understood the implications of Lyme disease from personal experience.

Four years before, in 2010, Dutch citizens had also delivered a message to Parliament, submitting a citizens' petition that legislators were legally obliged, by virtue of 71,000 signatures, to take up. Do something to monitor, prevent, test for, and treat Lyme disease, it implored. Barbara's email arrived at the petition's tipping point, the make-or-break moment for a movement seeking recognition. By then, a task force had met, hearings had been held, and committee reports had been issued. Advocates feared the battle would be lost in the muddled politics and polarized debate of Lyme disease. Barbara's electronic missive, which she posted on a Lyme website where it was received with alarm, would make Dutch history. Sent March 19, 2014, just before 11 p.m., it began, "As my last wish before I will end my life...."

Antibiotic Line in the Sand

Every year, some 25,000 people in the Netherlands are diagnosed with Lyme disease, a figure that quadrupled from 1994 to 2015. The majority of people are treated successfully with a course of antibiotics, typically doxycycline or, particularly for young children, amoxicillin. These success stories have been used to mold the public perception of Lyme disease as a benign illness, common to some geographical regions, the story goes, but eminently treatable. Missing from that view, which has been fostered by the medical establishment in America, is the other side of Lyme, the more insidious one. This is the one in which tick bites go unseen, tests fail, cases are misdiagnosed, and, of growing importance, illnesses are complicated by other pathogens carried by ticks. In 2001, a Columbia University team published a study of twenty children, ages eight to sixteen, whose parents had taken them to an average of four doctors, with some seeing seven physicians, before Lyme disease was confirmed. Those diagnostic delays came at a price. Eighteen months after treatment, the children suffered many more memory and attention problems than a healthy control group and had lower IQs. They had trouble retrieving words and were hypersensitive to light and sound. Ominously, eight of the children, or forty percent, had had suicidal thoughts, and two had made suicidal gestures.

These children were tested for cognitive not physical problems of pain and fatigue that are also common to people belatedly treated. They are but a small sample of what the Columbia researchers called chronic Lyme disease but is officially termed Post-Treatment Lyme Disease Syndrome. Whatever it is called, there is no question that significant numbers of people suffer ongoing symptoms from the bites of diseased ticks that go far beyond the cognitive issues of the Columbia children. How many, of all those infected, for how long, and how to treat, are unknown. ·

Consider patients in Holland, where Barbara Pronk lived. Of those 25,000 Dutch cases reported annually, 1,000 to 2,500 people were estimated in 2015 to suffer prolonged symptoms a year after treatment, including fatigue, pain, and problems with concentration. The figure is likely low because of how Lyme disease is reported in the Netherlands, with most people diagnosed by the classic Lyme rash—the easiest cases to flag and treat. Other studies suggest there are many more posttreatment sufferers. At Johns Hopkins University, sixty-three Lyme disease patients—"highly active and healthy" before infection—were treated with antibiotics and, because they were symptom-free just afterward, were thought to have been cured. But six months later, twenty-one of the patients, 36 percent, developed new symptoms of pain, fatigue, and memory problems. Lyme disease or not, many people suffer in such ways, especially as they age. But this group's rate was far higher, a challenge to assertions that posttreatment complaints "appear to be more related to the aches and pains of daily living rather than to either Lyme disease or a tick-borne coinfection." That wording comes from the 2006 medical guidelines that have governed Lyme disease care in the United States and, because of the influence of American research and medicine, in much of the world. There is considerable science to suggest that such guidance, proffered by the Infectious Diseases Society of America (IDSA) is little more than opinion and not based in fact.

For at least five years, the US Centers for Disease Control and Prevention (CDC) had officially stated, based on a review of the scientific literature, that 10 to 20 percent of treated Lyme patients exhibited symptoms six months or more after they were treated. But typical of the controversy that dogs Lyme disease, as I'll discuss in chapter 4, the agency changed that wording in 2015, saying instead that a "small" share of patients suffer ongoing problems. This is characteristic of a long-running effort to tamp down public fear of Lyme disease and disempower an unhappy patient community. The CDC, like medicine generally, does not officially recognize that Lyme disease may be "chronic," defined as the failure of common antibiotics to kill the infecting spirochete, leaving patients sick. And yet it cannot explain, as officials in Holland put it on a government website, "why some patients develop such complaints and why these persist."

I have heard the stories of many people who suffer long-term effects from the bite of a diseased tick. Many see confirmation that they have a chronic form of Lyme infection because they respond, to a point, to antibiotics, and relapse without them. Others had suffered from a complex array of tick-borne infections and been rejected for care by top-notch clinics and doctors. These patients—from the United States, Canada, European countries such as Sweden, the United Kingdom, the Netherlands, Germany, and Austria, and even Australia—often were diagnosed months or years after initial infection, giving the Lyme spirochete time to disperse throughout the body and cause a laundry list of symptoms. "A fork felt like a bowling ball," wrote New York City singer and songwriter Dana Parish, who, though treated early, was so overwhelmed by weakness she believed she would die. A Pennsylvania woman, Sherrill Franklin, who had founded and run a highly successful manufacturing company, suffered vertigo, tinnitus, and a thirty-pound weight loss. "I struggled to climb a set of stairs," she said. An Oxford University student, bitten by a tick in her first year, could once memorize Shakespeare plays; with Lyme, "my memory had all but shut down," Rachel Martin wrote in an essay describing her ordeal. Darren Fisher, from a remote town in Alberta, Canada, described daily headaches and debilitating back pain that $250,000 spent on treatments and care had not curtailed. His one ironic stroke of luck was when a radiologist certified brain damage enough to qualify him for long-term disability.

The experiences of these patients are the reason that Lyme disease should be feared. It is a stealth invader that is remarkably adept at evading the body's immune system and finding places to hide from both diagnostic tests and antibiotic treatment—concepts that are backed by science I will discuss in coming chapters. In its advanced stages, Lyme disease is an illness modern medicine is ill equipped to treat, let alone cure.

These patients tell remarkably similar diagnostic sagas to the one lived by Barbara Pronk. Symptoms were assessed. Tests were performed. Medical wisdom was checked. And patients were sent elsewhere for answers. A young man who grew up with my children on our country lane visited perhaps ten doctors when Lyme disease made it impossible for him to work as a landscaper. "After awhile," Richard Hargrove said of the physicians he'd seen, "I knew that look." A woman in Holland told me of going to the Czech Republic for her baby's Lyme disease treatment, after local physicians refused to care for the little boy. A woman from London saw eleven doctors in six specialties; after she tested positive, a rheumatologist confessed to knowing nothing about treating Lyme disease and sent her away. Indeed, in the face of unschooled doctors and uncertain treatment, patients go to great lengths for care. In Canada, they cross the border into New York State to see a particular doctor because Canadian physicians won't take them. The Oxford student sought care in the Netherlands and later the United States. A doctor in Washington, DC, Joseph Jemsek, has a world map with dozens of pins dotting the many countries from which his Lyme patients come. These pins represent people who may exhaust their life savings and mortgage their homes to seek care outside the standard medical menu that has failed them. Many others have no such resources to spend.

When turned away by doctors, patients with advanced cases are given various reasons, some valid, some in not so many words: You think you have Lyme disease but probably don't; it says so in the guidelines and the literature. You are depressed, anxious, mentally ill, misguided. You will take too much time. I do not know how. The insurance company won't pay for your care.

What these doctors surely do not say is, I do not want to lose my license or face disciplinary charges, as others have, for prescribing antibiotics for longer than medical protocols suggest, which is generally ten to twenty-eight days. Imagine this. We are talking not about prescription opioids that killed 28,000 people in the United States in 2014 and for which few doctors have been held responsible. We are not talking about pharmaceuticals with alarming toxic side effects—duly noted in scores of American television advertisements—that are readily dispensed after government-sanctioned risk-benefit calculations. We are talking antibiotics, inexpensive drugs that, while no doubt overused generally and plagued by issues of superbug resistance, are among the safest there are.

In 1997, a Boston University Medical Center physician named Sam Donta wrote a paper that began, "Two hundred seventy-seven patients with chronic Lyme disease were treated with tetracycline for 1 to 11 months (mean, 4 months); the outcomes for these patients were generally good. Overall, 20% of the patients were cured; 70% of the patients' conditions improved, and treatment failed for 10% of the patients." Two decades later, doctors like Donta are ostracized in the name of keeping patients safe from overuse of antibiotics. Yet there is little data showing the risk of "adverse events" tied to such treatment is any greater than the accepted risks of many other medications. Reports on the worst outcomes of prolonged antibiotic treatment for Lyme disease, namely serious infection and death, are few and focus almost exclusively on intravenous care. They are also used to instill fear in doctors who consider any kind of longer antibiotic treatment.

A brief report on the death of a thirty-year-old woman from a fungus in an intravenous catheter used to dispense antibiotics has been cited seventy times in the scientific literature. The short case report of the death of a fifty-two-year-old woman has been cited thirty-eight times; after ten weeks on IV treatment, she developed an all-too common infection from *Clostridium difficile*, which kills 14,000 Americans annually. These footnoted cases are used to bolster blanket dismissals that state, as one did, that long-term antibiotic treatment has "contrib-uted to injury and even deaths of patients." Such citations rarely say how many deaths. A list of studies posted by the CDC on the risk of prolonged Lyme disease treatment includes just six published papers in twenty-one years; four are single reports of deaths related to intravenous care. A fifth covered gall bladder complications, and a sixth focused on "unorthodox" treatments "marketed" to patients. This is hardly an extensive body of literature.

Antibiotics are not risk-free and should be used only in spare and appropriate measure. In a Lyme treatment trial published in 2003, four of fifty-five patients had severe side effects from antibiotics, including one allergic reaction and three intravenous-line infections. Holly Ahern, a New York microbiologist and scientific advisor to patient advocates, said, frankly, "I am not a fan. They cause collateral damage to our micro-biome." For one, overuse of the drugs can promote antibiotic resistance in patients to other infections. Yet while the risks of prolonged antibiotics are common to and acceptable for conditions such as acne, urinary tract infection, or tuberculosis, a line has been arbitrarily drawn on Lyme disease. And it is based on medical recommendations that have been raised to the level of dogma.

In the years since intravenous-line infections were reported in the 2003 Lyme trial, and in other trials as well, strides have been made in the safety of intravenous line care. The *Journal of Hospital Infections* pointed in 2011 to "huge success…in reducing catheter-related bloodstream infection." But, most significantly, Lyme physicians have seen patients improve in treatment regimens that often involve multiple, usually oral, antibiotics. They generally prescribe them along with drugs to treat other infections delivered by ticks; antimicrobial herbs, and sometimes, antibody-boosting therapy derived from human plasma. "The risks of oral antibiotics are almost negligible with a proper diet and adequate probiotic supplementation," said a leading Lyme disease physician in

New York State, Richard Horowitz, who sees IV antibiotics as helpful in neurological cases but "not viable" in the long term. "The risk however of not properly treating chronic Lyme is a life filled with pain and disabling symptoms."

Doctors audacious enough to take on long-term Lyme disease know that antibiotics have limited effect and are trying to figure out what works. For thirty years, Horowitz has been on this quest, convinced that prevailing treatment recommendations have led to a "dysfunctional" model of care. He reported significant success in 2016 when he combined antibiotics with, of all things, a sulfa drug used for leprosy, called Dapsone. Horowitz, whose findings were published in *Clinical & Experimental Dermatology Research*, had picked up on the success of test tube experiments at Johns Hopkins that found the drugs effective at killing spirochetes that had survived initial antibiotic treatment. This was a marriage of laboratory findings and real-world practice to find out what works in intractable cases, a kind of work that is, for Lyme disease, all-too scarce.

The line in the sand on Lyme-dispensed antibiotics means patients in the United States and other countries cannot get drugs their doctors might actually want to prescribe. An Austrian woman said she feared that her doctor might refuse her more antibiotics if her name was used in this book because it would potentially identify him. A Canadian man, Patrick Leech, convinced his doctor to grant two months of doxycycline after suffering facial palsy, joint pain, and muscle spasms; the pharmacist called the doctor, and he was denied. Lyme specialty physicians in many states have withstood licensing board investigations that centered on their use of antibiotics, dogging their practices for years, and leading other physicians to steer clear.

Instead of treating and judiciously retreating, doctors misdiagnose and refer patients elsewhere, underestimate the implications of Lyme disease, and avoid patients altogether. This happens especially in later cases, but in early ones too. Doctors have been told that medicine has Lyme disease figured out. "There's no evidence in North America of persistence of the spirochete after treatment," the chief author of the Lyme treatment guidelines, Dr. Gary Wormser, told me in an interview in 2012. "We've published on it, and we've looked. You show me a credible study in the U.S. that shows persistence of the organism after treatment." There is, indeed, such evidence, some published before I spoke with Wormser, who declined to speak for this book, and a good deal afterward. These studies, which I will discuss in chapters 4 and 7, document lingering spirochetes in antibiotic-treated animals and their survival after being dosed heavily with antibiotics in test tubes. Animal and test-tube studies are accepted means of testing hypotheses in many scientific endeavors—and for spurring further research in humans—but their findings have been rejected for Lyme disease. These published studies, however, demonstrate the significant shades of gray in the black-and-white picture of Lyme disease care as practiced in America. While antibiotics are by no means a cure, and intravenous antibiotics pose risks, there is evidence that additional courses can provide relief for lingering, persistent, intractable Lyme disease. And there are reasons to question the trials and studies that have found otherwise.

"Lonely Battle"

When Barbara Pronk's email arrived in inboxes, there was a furious scramble to save her. Her website post was taken down. Police were dispatched to the flat by the sea. Colleagues were horrified. One of them read the email in Australia, where she had been transferred, and instantly knew: Barbara, a serious, competent woman, a perfectionist at planning, was already dead. Barbara had indeed planned well. She had bought the medications she took online. She had rented the apartment for the purpose of suicide. She had picked out her coffin.

It matters little that Barbara Pronk, who died alone, may have contracted her illness from the bite of a tick in Florida or in her home country at some other time. The factors that led her own survival to clash with that of a tick are widely shared in many parts of the developed world.

After Barbara's death, the Dutch Parliament made a commitment to address Lyme disease. Three clinics were planned to treat the disease, though arguments have ensued on what treatments to use and for how long. Research money was made available by Parliament, enabling various experiments. Sheep were employed in forest patches to attract and remove ticks. Deer were fenced from areas to test the effect. Ticks were monitored in the dunes and heather. Climate change was measured. Surveys were done, finding that 31 percent of Lyme patients were infected in their own gardens. Said Willem Takken, the Dutch entomologist, "There is no nature area where you are safe."

For Barbara, Lyme disease was "a long tough battle, a lonely battle, a hopeless battle, a grueling battle." In her final note, she called her experience "degrading," the disease "not recognized and heavily underestimated." Barbara's experience—her maze of skeptical doctors and ineffective treatments, her intolerable pain—led her to believe that suicide was the "number 1" cause of Lyme death. She may, indeed, be right. A New Jersey psychiatrist who had long treated Lyme disease went through his case records in a study of patients with suicidal thoughts and actions. He found many who had attempted or considered suicide after a Lyme diagnosis who had never done so before, and a host of suicide risk factors among Lyme patients: "explosive anger, intrusive images, sudden mood swings, paranoia, dissociative episodes, hallucinations, disinhibition, panic disorder," and so on. When he calculated the frequency of these observations and applied them to the estimated population of chronically ill Lyme patients, the psychiatrist, Robert Bransfield, concluded there were perhaps 1,200 suicides in the United States each year. His paper, published in 2017 in the journal *Neuropsychiatric Disease and Treatment*, was the first to explore the association between Lyme disease and suicide and was far from a definitive tally of death. But it raised an important question surrounding a disease that is known to invade the mind, cause depression and cognitive problems and, moreover, send its victims for help in a system that offers little.

Around the time of Barbara Pronk's death, there were five Lyme disease suicides in the Netherlands. In one, a twenty-seven-year-old man from the central city of Lelystad, Jeroen Link, jumped from a rugged limestone cliff on the English Channel in Normandy, France. Jeroen had been an energetic, trouble-making, endearing youngster. He grew to a blonde and muscular adult who loved playing soccer with his friends despite struggling with Lyme disease since he was fifteen. He too had been told, in the common Dutch expression, that his malady was between his ears. When he told his parents what he intended to do, they were not surprised. "We talked, hugged, cried," his father, Johan, wrote on Facebook two months before Barbara's death. "What could I do? I was numb. Ideally I wanted to keep him with me, of course, but I also gave him peace and quiet. No more pain." Commented one woman in response, "Those of us that are afflicted by this, know exactly where this young man's mind was." Jeroen's suicide, covered in the Dutch press along with the others, was one more reason the Dutch Parliament took up the Lyme disease cause.

"Please imagine, your own partner, son, daughter, father or mother was affected by this monstrous disease," Barbara implored in her final note. "A healthy person has 1,000 wishes, a sick man only one." Barbara wasn't asking for her health. That was gone. She was asking for change.

CHAPTER 3:

An Ancient Bug Revives

The minuscule, ringlet-shaped organism known as *Borrelia burgdorferi*, the source of considerable human misery, has seen the world through the ages. Fifteen to twenty million years ago, a baby tick in the Cordillera Septentrional Mountains of the Dominican Republic sipped the blood of some prehistoric animal and was caught in a torrent of sap from a pine tree. The gut of that prehistoric tick, seen under a microscope in the twentieth century, would reveal wiggling corkscrew-shaped Lyme bacteria, preserved in motion like the ash-covered victims of Pompeii. A few million years later, in the Pleistocene Age, the bacterium that would be the driver of Lyme disease was caught under a global sheet of ice, survived a two-million-year sleep, and emerged 12,000 years ago when the ice retreated.

Nestled, like its forebears, in the alimentary canal of hard-bodied ticks, *Borrelia burgdorferi* and the ticks it inhabited thence began a northward migration saddled on birds or attached to roving vertebrates. Genetic markers tell us these post-Ice Age bugs moved swiftly from the southern United States to the Northeast, then the Midwest, galvanized much as they are today by a warming climate. Across the Atlantic, meantime, the bug was ensconced, among other places, in the Italian Alps. There, a bearded shepherd, nicknamed Ötzi, was shot in the shoulder with an arrow some 5,300 years ago, his forty-five-year-old body entombed in a glacier. A belated postmortem revealed this prehistoric murder victim to be the first known case of Lyme disease. Quite remarkably, he even carried a mushroom-like fungus with antibiotic properties, perhaps to quell the pain, to keep the infection in check.

The children of Lyme, Connecticut, from which the disease took its name, were not *B. burgdorferi*'s first young victims, as they most certainly were not the last. Some five hundred years before the birth of Christ, the Lyme pathogen lived inside the knees of prehistoric Tchefuncte Indian teenagers in Louisiana, whose buried bones show rheumatoid arthritis-like disfigurement since linked to *Borrelia*. That particular Lyme disease endemic subsided, scientists speculate, after the hunter-gatherers turned to agriculture, giving ticks less chance to bite. Yet the pathogen remained, infecting the tissue of colonial Americans and prospering amid the abundant forests, meadows, and deer of America and Europe in the 1700s. In 1774, a Rev. Dr. John Walker precisely described what is today believed to be quintessential Lyme disease on an island in Scotland's Hebrides. "This disease," he wrote, "arises from a worm lodged under the skin, that penetrates with exquisite pain, the interior parts of the limbs."

To this point in human history, the fortunes of *Borrelia burgdorferi* were rising. Then, in the 1800s, things changed. Deer were hunted for meat and fur. Agriculture slowly consumed the forests that sustained the mammals and, by extension, the ticks to which they were literally linked. By 1900, the North American deer population plummeted to almost zero, from perhaps six million in 1500. European stocks dwindled for the same reasons. For centuries, small mammals had harbored and nurtured the spirochete during the first two stages of a tick's life: as babies, called larvae, and as juveniles, or nymphs. But deer were key to the mature arachnid. The ungulates were the marital bed for the adult arachnids; they were the place where each succeeding generation of *Ixodes* ticks was made possible. "Deforestation of much of the Northeast during the 18th and 19th centuries resulted in the near total elimination of deer, and presumably also of deer [or blacklegged] ticks," wrote the first researchers to posit this theory, Alan Barbour and Durland Fish, in 1993. Gone were the deer. Gone were the ticks. Gone was *Borrelia burgdorferi*. Or at least, that is how the latest chapter in the on-again-off-again life of *B. burgdorferi* goes—a bug that natural history shows was not to be written off, and certainly not this time.

George Poinar Jr. is an Oregon entomologist and college professor who delicately dissects fossilized resin, huge yellow chunks of it, to unravel secrets on an epochal scale. It was Professor Poinar who found that tick in Dominican amber, its gut awash in what are officially termed "*Borrelia*-like" cells. (He used a strong microscope to identify the spirochetes; extraction would destroy the specimen.) That, however, wasn't by a long shot his oldest discovery of pathogen-packed ticks. Encased in another chunk of amber from Myanmar, half a world away and a hundred million years old, he found another set of cells, *Rickettsia*-like, inside the belly of a very old tick, the hard kind like *Ixodes*. As with *B. burgdorferi*, *Rickettsia rickettsii* emerged around the turn of the twenty-first century as a pathogen to watch. The cause of sometimes-fatal Rocky Mountain spotted fever, its numbers, though small, have soared in the United States since 2000 as ticks move and grow in number. Studying ancient ticks makes Poinar think they and the microbes they carry have aligned themselves over the ages in a compact that somehow works for both, a notion supported by genomic research.

"I feel that ticks were carrying and transmitting *Borrelia*-like organisms for millions of years," he told me. "A close association evolved between these two organisms that allowed the successful transfer of pathogens…back and forth from vertebrate [the animal] to vector [the tick] and vice versa. The longer two organisms have been associated with one another, the more finely tuned the association becomes." Poinar's research points to the most primitive origins of modern-day ticks and the cargo they haul. For all we know, *Borrelia* may have been in a tick that hitched a flight on a pterodactyl, so old and resilient is this remarkable, indefatigable, unfathomable organism.

A Bug of Many Talents

The story of how and why *Borrelia burgdorferi*, and with it Lyme disease, rather suddenly materialized in 1970s in the Northeast state of Connecticut is not simple. Nothing, indeed, is simple about the spirochete's mechanisms and methods, which are nothing short of a symphonic arrangement involving the guts of ticks; the blood of animals; and all the changes of temperature, viscosity, and chemistry therein. Nature is by definition complex, of course, but this ancient bug stands apart.

Remarkable? Like the tick that carries it, which can sense the breath of a potential meal from fifty feet, this spirochete has special talents. When it passes through the mouth of a tick to the animal or human the tick has bitten, *B. burgdorferi* stops producing a protein on its surface that helped it adhere to the tick's digestive tract (so as not to be swallowed whole). By thus "down-regulating," as it's called, the bacterium relinquishes a red flag that would otherwise signal its host to kill it. And when it passes into the human bloodstream, which for them is akin to a kayaker on the Colorado River after a raging storm, *B. burgdorferi* swings on tethers within blood vessels, planting anchors along the way to steady and slow its movement. That's not all by a mile.

Indefatigable? Millions of years of evolution have blessed *B. burgdorferi* with a powerful, and hidden, propulsion mechanism. Its flagellum—that tiny mechanism that helps a bacterium swim—is internal, shielded by a membrane that lets it bypass alarm bells normally set off in any self-respecting immune system. This structure allows it, moreover, to swim Olympian-like through fluids that would be the death of many bacteria and to penetrate tough joint capsules, the pericardium around the heart, and meninges that encase the brain, something few bacteria can do. Tara Moriarty, a *Borrelia* researcher at the University of Toronto, likened it to a fast-moving snake, a "flat wave," skinny at one end and as comfortable in liquid as in solid. Coiled as spirochetes are, they look under microscopy like so many tiny, whirring drills.

Unfathomable? A team of scientists that mapped *Borrelia burgdorferi's* genome in 1997 found an organism unlike any other disease-causing pathogen known to science. For one, it had "few, if any, recognizable genes involved in virulence," the researchers wrote, while lacking the mechanics to make the basic building blocks of life. It simply takes what it needs, "apparently scavenging these necessities from the host," they said. At the time, other scientists, writing in the journal *Nature*, said *B. burgdorferi's* newly unveiled genome "challenged ideas of what a bacterial chromosome is," with "an almost bewildering array" of genes unique among bacteria. More perplexing, it had qualities that suggested it could change its genetic signature—"to duplicate genes and rearrange them without much cost or damage." Left unexplained, but perhaps hinted at, was why the organism could survive in so many places— from the salivary glands of ticks, to the joints of horses and mice, to the hearts and minds of humans—and, moreover, how it could persist in and migrate throughout its hosts. Those findings would come later.

To be sure, even after deer herds plummeted to near zero by 1900 and forests were clear-cut, *Borrelia burgdorferi* remained, unnamed, misunderstood, infecting ticks and people randomly. It was building slowly, in Europe in particular. Arvid Afzelius was perhaps the first physician to describe the Lyme rash, at a dermatology meeting in 1909 in Stockholm, Sweden, a country from which many future cases would come. A year later, a Viennese doctor reported three more rashes, and in 1911, a Swiss physician noted another. In 1913, an Austrian dermatologist and virologist Benjamin Lipschütz christened the radiating red mark "erythema chronicum migrans." Over the next two decades more Lyme-like cases were reported, some with serious symptoms of paralysis, meningitis, and arthritic pain, many involving children, and all in Europe. In 1922, a fifty-eight-year-old sheep farmer in France was bitten by a tick on his left buttock, developed a plum-sized red rash, and is considered the first case of Lyme radiculoneuritis, which affects the nerve roots and causes numbness, tingling, and stabbing pain. According to the 2012 book "Aspects of Lyme Borreliosis," Lyme disease was presumed in reports of children with colorful, radiating rashes in 1915 and 1920; in a thirty-four-year-old man with severe pain and fatigue in 1934; in twenty-three patients with facial palsy, severe headache, and other neurological problems in Germany in 1941. By 1943, a Swedish doctor presented a paper on 142 cases, his own and from the literature.

Throughout the cataloging of European cases, the organism was undoubtedly in North America—2 of 280 mice collected from 1870 to 1919 in coastal Massachusetts would later test positive for Lyme disease. But human cases were slow to emerge, or at least to be recognized. In 1970, a Wisconsin physician, who had been tick bitten in 1968 while grouse hunting, was fortunate enough to see an astute Marquette School of Medicine dermatologist. The skin doctor remembered something from his studies: a report from 1949, in which a Swedish doctor had used penicillin on a radiating red rash. Rudolph Scrimenti prescribed antibiotics, successfully, and the hunter became the first official Lyme disease case in the United States. But in 1900s America—as in parts of it now—Lyme disease was likely mistakenly diagnosed as the flu, allergies, arthritis, signs of aging, melancholia, psychosis, neuralgia, and any of a long list of maladies that it mimics. Later it would be mistaken for other ailments as well, like fibromyalgia and multiple sclerosis.

New York State's Long Island hugs the Atlantic coast for 118 miles from Brooklyn in the west to the dunes and bluffs of Montauk Point in the east. Here, in the region where Lyme would explode in the last quarter of the twentieth century, 136 ticks were collected in the 1940s from two oceanside locations, Montauk Point and a slip of land on the South Fork known as Hither Hills. When the specimens were tested in 1990, thirteen of them, about 10 percent, were found to harbor the Lyme bug's genetic fingerprint. Those ticks made their home just thirty-two miles, as the crow flies, from Lyme, Connecticut. There, in 1975, two mothers would demand answers when a raft of children came down with swollen knees and arthritis-like symptoms that were later attributed to the bites of ticks and, more specifically, to a pathogen within the ticks. Yale School of Medicine became involved, and ticks were identified as the carrier, or vector, of infection.

The pathogen itself would be identified in 1981 by Willy Burgdorfer, a researcher at a US government facility called Rocky Mountain Laboratories, in Hamilton, Montana: *Borrelia burgdorferi*. At last, after eons of infection, sightings from Central America to the Alps, and much human suffering, the bug had a name. Europeans, too, recognized this illness. In Germany and Austria in 1996, twenty-five *Ixodes ricinus* ticks dating from the 1880s were dug out of institutional archives. Two tested positive for species of *Borrelia* linked today to Lyme disease. "Residents of Europe have been exposed to diverse Lyme disease spirochetes at least since 1884," researchers reported, "concurrent with the oldest record of apparent human infection." The oldest record, perhaps, but certainly not the oldest case.

A Switch, Flipped

For all the scattered human cases in 1900s Europe, Lyme disease was a long but a minor medical mystery, a small public health concern. Something happened, something coalesced, around the mid- to late-1900s to animate this millennia-old tick-borne malady. *Borrelia burgdorferi*-laced ticks were was no longer an atypical risk in a rare forest or field. Lyme disease was no longer the novelty of early 1900s Sweden for the dramatic red flag it bestowed on its victims. This disease grew, in the last quarter of the twentieth century, from cases measured by the handful to an epidemic infecting millions worldwide, one belatedly recognized and woefully underappreciated. To be sure, the United States began officially counting cases in 1991, meaning that doctors were mandated to report them to health authorities. Hence, some of the rise over the last quarter century can be attributed to awareness. Was this a reporting phenomenon, since we know the bug had long been in the environment? Or was this a sudden explosion that drowned out *B. burgdorferi's* background noise? Could an epidemic and a phenomenon like this have been overlooked for long? This wasn't just more cases, however; this was more ticks.

In Sweden, where some of the oldest tick data are kept, maps from before 1980 show the arachnids confined mainly to a broad but tight arc around the Stockholm archipelago; by the mid-1990s, the ticks gravitated outward into central Sweden and north along the Baltic coast. In 1995 and in 2006, Dutch authorities conducted national surveys in which they asked residents simply if they had been bitten by ticks. The per capita rate of tick bites jumped 75 percent in a decade—"a progressive threat to public health," officials called it. *Ixodes* ticks have surged across the continental United States, moving from less than a third of counties in 1998 to half in 2015. In the state of Iowa, *Ixodes* ticks were seen in nine counties in 1990; by 2013, they had turned up in seventy-two counties, where there were clear signs this was more than a counting phenomenon. While tick numbers rose, so did the proportion infected with the Lyme spirochete, hitting a peak of 24 percent in 2013, the final year of the study. That is triple the rate of 1998.

In upstate New York, where I live, ticks are a frequent topic of conversation among long-time residents. "Do you remember seeing them 30 years ago?" a friend asked me. We agreed; we did not. I recall a large black tick in the fold of my ear after jogging through brush one day in the 1980s; this was an exotic creature indeed. By the early 2000s, all that had changed. After a late spring walk in a field around 2010, a helpful friend and I removed forty blacklegged adult ticks from my two dogs before we stopped counting. We were aghast. By then, we knew the upshot.

What flipped the switch to make this happen, in that field and in our world, were gradual, inexorable changes, long in the making, on the ground and in the atmosphere. These changes were prompted by how and where people lived, what animals they fostered and which ones they displaced, how they got around and warmed and cooled themselves, how they were fed. Human influence had become so powerful, so vast, that it exerted control over the mechanisms of weather, temperature, global ecology, and contagion. Lyme is listed among the giants of disease that have been and will be fueled by global climate change, including cholera, dengue, and malaria. The difference is that those are long-known agents of misery, while Lyme is new. The US Global Change Research Program predicted with "high confidence" in 2016 that a warmer world would prompt ticks carrying disease to come out earlier in spring and to move generally northward. With the same confidence, it forecasted more mosquito-borne diseases, like West Nile, a prediction outdated only months later. That was when the Zika virus—a major threat for its potential for sexual transmission and brain damage in utero—burst on the scene. Others, to be sure, lie in wait.

But Lyme disease was the first to arrive, to blossom and thrive in a wide swath of earthly territory, in the era of climate change. The US government tracks case counts as a measure of warming, along with wildfires and heat-related deaths. Lyme disease is the biggest vector-borne disease in the United States, and, when undercounting of cases is factored in, the nation's second-leading infectious disease as well. Why Lyme disease exploded precisely when it did is not fully understood. Whether climate change prompted the epidemic or merely helped it along is an open question. But of this there is little doubt: the hand of humankind shaped this epidemic, aiding and abetting the many forces that went into creating it.

Sarah Randolph is a zoologist and Oxford University professor emeritus with close-cropped hair and a reputation that is legendary in the world of tick study. She has strong opinions on Lyme disease—"the go-to cause for any malaise," she thinks—and a fervor for the scientific study of eight-legged creatures that bite. The "beauty" of ticks, she has written, as only a zealot can, "lies in molecular detail." She marvels at the ingenuity of scientists who unravel their mysteries, from counting ticks on mountains to picking them off the chins of wriggling, and very unhappy, garden dormice. "Indeed fantastic!" is how she described this scientific pursuit to me. Professor Randolph's prime focus is *Ixodes ricinus,* the castor bean tick, and its potential to impart tick-borne encephalitis. The world needs people like her, who love the workings and wonders of ticks, though she has been a thorn in the side of a Lyme researcher or two.

In 2010, Randolph set out to study why cases soared in three former Soviet-bloc countries of tick-borne encephalitis (TBE), a viral illness prominent in Russia and Eastern Europe. TBE can cause seizures, meningitis, and problems of thinking and memory and kills 1 to 2 percent of those infected; it is far less prevalent than Lyme disease, with perhaps 2,500 cases in Europe yearly. In 2009, cases of the disease went up in eleven of fourteen European countries, with particularly startling increases, of 45 to 91 percent, in Poland, Lithuania, and Latvia. Was this surge in disease driven by more ticks, more deer, a warming climate? No, Randolph and a colleague concluded. Instead, as incomes dropped, residents took to forests that were tick-infested to forage for mushrooms and berries or to cut wood, long-standing practices among impoverished Eastern Europeans. At the same time, the cost of vaccination against the disease—a prime tool that sharply limits cases—soared

40 percent in Lithuania, the hardest hit country with the lowest inocu-
lation rate. This particular outbreak, Randolph maintained, was driven
by poverty, unemployment, and hunger.

Randolph's paper is one of several, cited scores of times, in which she
argues that social and economic factors drive disease by putting people
in the paths of ticks. She disdains buzzwords like biodiversity and is a
determined naysayer of climate change as a driving factor in tick-borne
disease. "Human behavioral shifts...after the fall of the Berlin wall were
far more significant than climate change," she told me.

It's the Deer

On the other side of the Atlantic Ocean, another kind of *Ixodes* tick, the
blacklegged tick, lives well and large in the gardens, carved-up forests,
and hilly trails of Dutchess County, ninety minutes north of New York
City. This tick is *Ixodes scapularis*, the prime mover of Lyme disease
there, and it is the life's work of another zoologist, Richard Ostfeld of
the Cary Institute of Ecosystem Studies in Millbrook, New York. In
1991, Ostfeld stepped into the deciduous woods of Dutchess County
in an effort to understand a plague of bug-eyed, hairy caterpillars that
had defoliated vast tracts of forest. Those gypsy moth caterpillars, so
numerous that poop rained audibly upon the forest floor, led Ostfeld to
a discovery that would chart his course for the next quarter century. He
found so many acorns that he slipped and slid on boots as if they were
lined with ball bearings. The acorns, it developed, begat a bumper crop
of mice. The mice, in turn, decimated the gypsy moths, feasting on the
delectable shell-encased stage between crawling and flying known as the
pupa. The caterpillar plague was over.

But it was what Ostfeld saw on the noses, cheeks, and ears of the
mice that predicted both his future and that of an epidemic. On each
mouse, there were twenty, sometimes thirty, newly hatched larval
ticks, feeding on the single best mammal from which to pick up the

pathogen that causes Lyme disease. Ostfeld was on a course to unravel the complex ecology of *Ixodes scapularis, Borrelia burgdorferi,* and *Peromyscus leucopus*—the tick, bug, and mouse that make up the troika of Lyme disease.

Ostfeld, a youthful sixty-something, is the senior scientist at Cary, an institute with unmowed grasses and low natural buildings that could double as a retreat center in the woods. Carved into six invisible grids, the Cary property—its flora and fauna—is also Ostfeld's window into the life of *Ixodes* ticks. Ostfeld measures how many larval, nymph, and adult ticks there are in the forest and how many are infected with Lyme disease. He counts the number of ticks on shrews, chipmunks, voles, mice, and birds throughout tick season and the proportion that are infected. He looks for, and has found, other pathogens in these ticks, counting coinfections as an emerging and troubling trend in the land of ticks. He has sometimes captured an exotic or invading species, like the two Gulf Coast ticks he saw in the summer of 2016 that gave him something to watch. He has figured out which animals are "competent reservoirs" of infection, namely those likely to nurse young ticks with *Borrelia*-laced blood. Bite a mouse—a highly competent reservoir—and a tick is likely to be infected with the Lyme bacterium 90 percent of the time. Bite a deer, considered "incompetent," and the tick will be infected less than 10 percent of the time.

In 2000, Ostfeld and his long-time collaborator, Felicia Keesing, introduced a groundbreaking idea into the science of disease ecology. Until then, ecosystems rich in diversity had been embraced for two reasons. They offered the promise of undiscovered substances from which to make future pharmaceuticals, and they supported animal populations key to the study of human disease. Now, Ostfeld and Keesing went one step further. A forest alive with a diverse array of wildlife, one species keeping another in check, they posited, could provide natural protection against disease. On the other hand, an ecosystem starved for diversity, the theory held, could be one in which pathogens bloomed.

The scientists' "conceptual model," their prima facie case of a pathogen enabled by a disordered, unbalanced environment, was Lyme disease. Devoid of predators and competitors, white-footed mice thrived in the "isolated woodlots and urbanized landscapes" of a postindustrial world, they wrote. In these diminished ecosystems, ticks feasted off the blood of the single best mammal to infect them with Lyme disease spirochetes: mice. Find an environment where mice are kept in check, one where ticks feed on animals that are not so chock full of Lyme disease spirochetes, and you will find less Lyme disease. Ostfeld and Keesing called this the "dilution effect." The pathogen is still there in nature, as it has been for eons, but it is watered down in its ability to circulate between host and tick and, hence, to proliferate. More foxes, for example, could mean fewer mice, which could lead to fewer infected ticks. So compelling was this theory that the Ostfeld-Keesing article, in the journal *Conservation Biology*, has been cited some five hundred times in the seventeen years since its publication.

Sarah Randolph does not adhere to the idea that natural forces working in harmony are, by definition, a solution for tick-borne disease. Biodiversity is a "mantra,"—"diversity protects against infection risk"— that, she has written, is "doomed to fail." Both esteemed researchers in parasitology, Ostfeld and Randolph have, in the arcane yet amazingly feisty way of scientists, disagreed strongly in print, in a debate that hits at the core of Lyme disease theory. Forget warm and fuzzy ideas on diversity, Randolph and a coauthor argued in a pointed critique of the Ostfeld theory. Forget climate change, what they called "the other environmental bête noir" of Lyme disease. These were merely "powerful levers for attracting funding from major agencies." Instead, blame deer: "Increased deer abundance results in increased tick populations," wrote Randolph et al., citing a good bit of published science that agreed.

In a book called *Lyme Disease*, published in 2011, Ostfeld devoted three chapters to what he saw as pat explanations for the spread of tick-borne disease. They are the following: "It's the deer." "It's the mice." "It's the weather." Each is important, he maintained. But each, in particular deer, has been oversold. For decades, he argued, efforts have been made to tame the rise of deer in the belief that ticks would be vanquished as well. Deer were, after all, the favored host of mating, egg-laying adult ticks. No sex, no babies, no problem. In 1982, 70 percent of deer were shot on Great Island, a six-hundred-acre landmass off the coast of Massachusetts. Mice, nonetheless, were found afterward to carry the same number of ticks as before, with some indications there were even more. Researchers had better success on Monhegan Island, off the coast of Maine, where all deer were eliminated in 1999 and ticks became scarce. And after 90 percent of deer were culled in a tract of 112 homes called Mumford Cove in Groton, Connecticut, in 2000, both ticks and Lyme disease cases dropped sharply. By 2011, however, researchers took another look at the data and reported that they "did not find a statistically significant effect of the deer hunt." Beyond this, deer travel. Controlling them on a mainland is much tougher than on an island, even if it could lessen tick populations.

The widespread contention that deer drive Lyme disease is, for Ostfeld, a worrisome myth. Killing or culling them could backfire and make things worse because deer also have one very big plus in their column. They generally do not infect ticks with Lyme disease. The nymph that feeds on a deer has a one in ten chance of becoming infected; the nymph that feeds on a mouse is infected nine times out of ten. Hence deer might actually dilute *Borrelia's* prevalence in the environment by feeding nymph ticks, which then molt into uninfected adults. What if deer were replaced, in the hierarchy of tick hosts, Ostfeld wonders, with another mammal that was better at transmitting the pathogen? Chipmunks, shrews, squirrels, even raccoons, all are better at imparting Lyme spirochetes to ticks than deer. Moreover, and this is key, it takes very few deer to support a great multitude of ticks, as the Great Island experiment found, each female laying a couple of thousand eggs at a clip. Eradicating or reducing millions of ungulates on a grand scale is impractical if not socially and politically impossible. And it also may not be wise.

It's the Weather

In my research for this book, I was fascinated by reports of Lyme-toting ticks in strange places. They moved up mountains in the Czech Republic and in Bosnia and Herzegovina. They were found above the 70th parallel north in Finland, nestled in the feathers of razorbills and puffins. One dropped off a dog in the Northwest Territory of Canada and prompted an alarming report on the TV news. These tick reports, from what one researcher called the "climatic edges of tick distributions," aren't terribly significant. After all, few people live on the tops of mountains or in Finnish Arctic. These reports are, however, indicative of a migration. Ticks are moving to places where people do live as well as to the fringes of human habitation. This movement can neither be stopped nor denied, and it aligns quite exquisitely with, well, the weather.

Consider a few places that bear witness to rises in temperature, ticks, and disease:

• **Norway:** A research team compared data on tick prevalence from the late 1930s and early 1940s to the early 2000s. Writing in 2001, the team said *Ixodes ricinus* was no longer limited to coastal areas as it long had been. Instead, it "is now found in inland and mountainous areas of South and Central Norway, and has expanded its northern distribution limit" by about 250 miles. The new tick homes include municipalities where many people live.

- **Sweden:** Researchers from Stockholm University and the University of California, Berkeley, charted the migration of ticks from the early 1980s to mid-1990s against changes in temperature. As the number of winter days below minus 12 degrees Celsius (10.4 degrees Fahrenheit) declined, ticks emerged in a consistent south-to-north pattern. The 1990s warming "is probably one of the primary reasons for the observed increase of density and geographic range of *I. ricinus* ticks," they reported in *Environmental Health Perspectives.*

- **Russia:** Using a prized set of tick observations over a thirty-five-year period, Russian researchers reconstructed the recent natural history of *Ixodes* ticks in the forests and floodplains of Tula, 120 miles south of Moscow. For the first fourteen years starting in 1977, the scientists saw few ticks at all. In the twenty-one years through 2011, however, they saw the population grow exponentially, what they called a "manifold increase" that they tied to "climate and anthropogenic changes."

- **The Arctic:** Venturing to a remote corner of the Siberian Artic called Verkhoyansk, a group of French and Russian researchers found the first evidence of Lyme disease, reported in 2016. With 10 percent of seventy-seven residents testing positive for Lyme pathogen antibodies—meaning they had been exposed to the spirochete but weren't necessarily sick—the researchers called the disease "a major health threat for people dwelling, sporting, or working" in this cold clime.

- **China:** Lyme disease, first reported with a smattering of cases in 1985, had been confirmed by 2013 in twenty-nine provinces and municipalities, according to researchers in Guangdong province. The numbers: three million bites and 30,000 Lyme disease cases in four provinces alone. "Continuous reports of emerging tick-borne disease cases…demonstrate the rise of these diseases throughout

China," their study said. Temperatures across China, meantime, are projected to rise 1.3 to 2.1 degrees Celsius by 2020, and 2.3 to 3.3 degrees Celsius by 2050, a trend that fosters vector-borne disease in China and elsewhere.

On the long list of Lyme disease theories that are open to debate, the role played by climate change ranks relatively low. As the world burns, *Ixodes* ticks are enjoying a population explosion and territorial expansion as never before; on that there is broad agreement. In 2006, a World Health Organization report made the link between a warming world and more of the castor bean ticks that carry Lyme disease in Europe. "Increases in density and...distribution of *I. ricinus* into higher altitudes and latitudes," two Scandinavian researchers wrote for the WHO, "are correlated to changes in local climate." Among the implications that can be argued is whether some places may ultimately get too hot or too dry to support ticks and the pathogens they bear. This is potentially a golden upside of climate change, of which there are few. For now, the story is more ticks in more places.

Whether we are poor in a Baltic state or middle class in Holland, a peasant in north China or a farmer in the American heartland, many of us are coming into closer contact with ticks. For this, we can blame patterns of development, reforestation, resurgence of deer, a plague of mice, the popularity of urban parks and suburban trails, and, as Randolph argues, economic forces that send peasants into forests for food. But these societal, recreational, and socioeconomic factors are just part of the explanation of tick-borne disease. They do not fully explain why ticks, those tiny titans from the dinosaur age, are moving to our doorsteps in the first place.

CHAPTER 4:

A Disease, Minimized

William James was a physician and psychologist who died in 1910 when he was 68, leaving a legacy of thought on the intersection of human consciousness and mental health known as functional psychology. Ranked fourteenth on a list of influential twentieth-century psychologists (after Skinner, Piaget, and Freud), he once said, "The greatest discovery of my generation is that a human being can alter his life by altering his attitudes." Aligned with that philosophy, James and others of his time suffered from what was then called "neurasthenia," a vaguely defined syndrome of fatigue, headaches, muscle pain, insomnia, and other symptoms. Its cause was thought to be the frenetic nature of late-nineteenth century America, with its growing cities and changing means of production, communication, and transportation. James popularized another name for this amorphous disease, rooted as it was in human response to circumstance rather than in documented pathology. He called it *Americanitis.*

In 2015, a group of infectious disease physician-researchers, as influential on Lyme disease as James was in the field of psychology, published the findings of a study of patients who had been bitten by ticks, infected with Lyme disease, and treated promptly with antibiotics. They were then tracked for eleven to twenty years. An average of fifteen years later, the study found the group doing quite well overall—just 4.7 percent had lingering pain and fatigue, symptoms of what has been called Post-Treatment Lyme Disease Syndrome, or PTLDS.

To that point, there had been many studies of how patients fared after they were given standard short courses of antibiotics for Lyme disease. Most research had found that significant shares of treated patients, 10 to 20 percent generally but as high as 60 percent for some patient groups, suffered fatigue, pain, memory problems, depression, and other issues even after supposedly curative treatment. But this single study—limited to a privileged set of early treated patients—would be used to redefine the dimensions of post-Lyme syndrome.

After tracking patients for an average of fifteen years, the eleven study authors essentially found little to worry about. Posttreatment Lyme, sometimes called chronic Lyme, was a problem for "only 6" of 128 patients or fewer than one in twenty, said the study. In an editorial that accompanied the study, which appeared in the journal *Clinical Infectious Diseases*, Paul Auwaerter, an infectious disease physician, said the new study "should help allay the common distress that Lyme disease is routinely life-altering for the long term." Today's chronic Lyme disease sufferers, he went on, citing James, were yesterday's neurasthenics, for which Lyme disease was "a universal explanation for fatiguing conditions." The editorial took pains to let Lyme patients down easy, noting James' contention that these psychosomatic sufferers were of "intelligent nature" with "intense yearning to try to label and understand" why they became suddenly ill, bedridden, "abandoned by family and friends." But, based on this study, Auwaerter attributed their maladies not to chronic, unresolved infection. Instead, he said, they suffered Americanitis.

In one study and one editorial, the most significant controversy over Lyme disease was being put to rest. For two decades, the Infectious

Diseases Society of America, or IDSA, has worked earnestly toward that end, having concluded early in its emergence that Lyme disease was eminently curable. But to the society's dismay, its worldview of Lyme has been tempered by a giant, well-informed, and pesky coterie of patients—tens of thousands of them in many countries—who fervently disagree. The IDSA side sees single courses of antibiotics as highly effective at curing the disease: the medications kill the Lyme disease spirochete in vivo, namely in people. Period. The other side—from ill patients who were told they were imagining things, to a few hundred marginalized but well-intentioned doctors, to researchers at selected prestigious universities—has challenged those assertions and amassed a convincing body of science to justify another look.

Following the publication of the Americanitis study, the debate over chronic or posttreatment Lyme disease took a curious, but perhaps predictable, turn. For at least five years, the US Centers for Disease Control and Prevention had asserted on its website that 10 to 20 percent of people who were treated for Lyme disease "will have lingering symptoms of fatigue, pain, or joint and muscle aches," sometimes "for more than 6 months." That amounted to 38,000 to 76,000 posttreatment sufferers of the 380,000 Americans estimated to be infected in 2015 alone. Those are huge numbers of potentially impaired people.

For years, the CDC's 10 to 20 percent figure, which was based on patients who had been treated early in the disease, was seen by Lyme patients as an acknowledgment that their suffering was real. Such numbers, however, did not comport with the view of Lyme disease fostered by the Infectious Diseases Society's treatment guidelines for Lyme disease. First issued in 2000 and updated in 2006, the guidelines maintained the tick-borne malady responded well to antibiotics—which it does early on—with few long-lasting problems, which is open to question. Those lingering problems, the guidelines state, could usually be attributed not to sickness but to "the aches and pains of daily living." Americanitis.

Six months after the posttreatment study was published, the CDC opted to change its stance on the after-effects of Lyme disease. As of June 2015, its website declared that "a small percentage" of patients, rather than 10 to 20 percent, might suffer continuing symptoms after treatment. The agency made no announcement of the adjustment, nor was it subjected to peer review, which is required for significant changes under a federal law designed to ensure "the quality, objectivity, utility, and integrity of information...disseminated by Federal agencies."

Forty years into the epidemic, in fact, the CDC had not conducted a single such review of its policies on Lyme disease, the nation's leading vector-borne disease and certainly the most controversial. Instead, it quietly made this change "based on new information," a CDC spokesperson wrote to me, naming the posttreatment study, "indicating a frequency of <5%." Curiously, the agency did not cite the new study on its web page, leaving in the reference from which the previous figure came.

Define "Small"

I relate this rather obtuse example of a bureaucracy at work because it is key to understanding Lyme disease policy in America. This change, in a critical government posting, grows out of a tight alliance between the CDC and the crafters of the IDSA's Lyme-treatment guidelines. Three of eleven authors on the posttreatment study wrote the Lyme guidelines, and three more are based with them at New York Medical College, the home of officially accepted Lyme dogma in America. That view, embraced by a government agency with worldwide influence, is open to question.

Here, as a case in point, are the caveats that were ignored in the CDC's decision to redefine the population of posttreatment sufferers as "small."

- The study's outcomes were measured an average of fifteen years after treatment—even though posttreatment Lyme syndrome is defined on the CDC website as symptoms that "can last for more than 6 months." In addition to the six patients considered in the study to have the syndrome, eight others—11 percent in all, rather than 5 percent—had symptoms three to six years after treatment. What about them? Even if 5 percent was accurate, that would amount to perhaps 17,000 patients *annually* who would still have symptoms after fifteen years. Is that small?

- More than half of the study's Lyme disease patients, 155, dropped out along the way, and these patients were sicker—with on average four major symptoms compared to three among those who remained. Were they unhappy with the traditional treatments they were offered? Did they go elsewhere? How "this potential selection bias might have affected results is unclear," Auwaerter's editorial acknowledged.

- But most significant, and at the heart of the Lyme debate, was that everyone in the study had been lucky enough to have been treated soon after infection—within a week of symptom onset, a golden time in the life of a Lyme infection when anything is possible.

Indeed, early treatment is key to assuring that *B. burgdorferi* spirochetes are stopped in their tracks, before they move to the brain, the eyes, the nervous system, the joints, to places that complicate treatment and confound tests. The CDC left out all of those relevant details in its blanket statement on the after-effects of treated Lyme disease, part of a misleading picture of Lyme disease as it is officially drawn.

In June of 2017, the CDC released a report on complications of long-term intravenous therapy for Lyme disease. Five patients in a three-year period, the report said, had suffered serious infections from the intravenous lines, including one who died. These incidents were undoubtedly concerning. But the report was as significant for what was left out. Aside from failing to balance the risk of treatment against its benefit, the agency's report did not say how it had compiled its cases. For this, CDC officials solicited stories on bad outcomes involving Lyme patients from Infectious Diseases Society doctors with whom the agency has long been aligned. The stories were meant to cite the failures of physicians who believe that Lyme disease can be chronic and who do not follow CDC-supported IDSA dictates. The medical association that represents those physicians, the International Lyme and Associated Diseases Society, was not contacted for the study, nor were key members I queried. These cases were little more than cherry-picked anecdotes, provided by the side with which the CDC agreed. At the time, a CDC spokesperson told me that no systematic look at the risks of intravenous antibiotic treatment for Lyme disease had ever been done. This surely wasn't it.

In five years of writing about Lyme disease, I have found a medical landscape that is breathtakingly controversial and, in many ways, dysfunctional, one characterized less by warring sides than by parallel universes. The side taken by the CDC strenuously and dogmatically proffers a view of the disease that is straightforward and unambiguous, solidified in 2001 by two major treatment trials and precious few since. The other sees a disease complicated by poor tests, a bafflingly complex and crafty bacterium, and the influence of related tick-borne infections, like babesiosis and bartonellosis, and a variety of related *B. burgdorferi* species and strains that may be missed by classic tests. Often denied government grants, this side has conducted research experiments with the help of nonprofit entities, challenging the status quo.

Unlike almost any other disease in America, Lyme disease care is a closely controlled construct of one American medical group—the IDSA—and one government agency, the CDC. Doctors who have opted to treat outside the mainstream have faced the terrifying prospect and reality of censure or suspension by licensing boards, here and in other countries. In England, France, and the Netherlands, where I visited; in Australia, Austria, Canada, Germany, Ireland, Sweden, and in other countries from which I interviewed patients and doctors, US guidelines on Lyme disease treatment dictate care. In Sweden in early 2017, a physician named Kenneth Sandström lost his license for using too many antibiotics to treat people with tick-borne infections, even though many attested to getting better under his care and 6,700 signed a petition in support.

Sandström and other doctors do this for Lyme patients, I found in dozens of interviews, because the drugs made patients better, though not always well. Lacking an alternative, antibiotics, augmented with herbal antimicrobials and treatments for coinfections, allow some people with advanced Lyme disease to work, raise children, and exist in a relatively pain-free state. Monica White, forty-seven, of Poncha Springs, Colorado, had been incapacitated by Lyme disease and three other tick-borne infections for several years when she began intravenous antibiotic treatment around 2013. "If this is taken off the table for me," said the mother and disabled forest biologist in 2017, "I don't have any other resort to keep me functioning." At a support group meeting in a local pub in Winchester, England, a woman in her fifties told me, as others have, "All I know is if I take enough antibiotics, I can have a reasonable quality of life. As soon as I stop, the symptoms come back in a matter of weeks." Like the vast majority of advanced Lyme patients, she takes the antibiotics orally and pays out of pocket. In treatment trials involving three-month courses of antibiotics, 9 to 46 percent of patients reported improvement from placebo drugs, which may in part explain why patients on longer-terms say they feel better. Those trials, however, generally showed greater, sometimes far greater, improvement in patients on antibiotics. These benefits were deemed statistically insignificant, temporary, or overshadowed by side effects. Reassessments of the trials, as I'll discuss later, have suggested their designs were poor and benefits underestimated.

Over the last quarter century, about two-dozen researcher-physicians—primarily the authors of the Infectious Diseases Society's treatment guidelines—have dominated Lyme disease care in the United States and around the world, a dynamic that has dramatically limited physician freedom. They have done this, with the support of major medical journals and government agencies in the United States and through a series of studies, much of it adequate if limited science. Their research findings, often repeated in drumbeat fashion in multiple journals, have defined and minimized Lyme disease by relying on successes in early cases, emphasizing the risks of longer treatment, and dismissing the reality of patient suffering. They have left little room for debate, even while they acknowledge the potential limits and biases of their research.

The Myths of Lyme

Coinciding with the dawn of open-access publishing, indeed perhaps because it provided a ready platform that is seen as cheapening its value, articles that have pushed back on traditional views of Lyme disease—particularly those suggesting evidence of the pathogen's ability to survive antibiotic treatment—have been dismissed as flawed or inconclusive. In Lyme disease, the natural process of scientific self-correction has been stymied, the battle for control less a war than a decades-old standoff. Over time, a tapestry of myths has been woven from sturdy but not incontrovertible science, creating an enduring and false likeness of Lyme disease as a problem without urgency. As a result, cases of Lyme disease have been missed, doctors have avoided caring for people with complicated presentations, and patients—many of them children—have suffered enormously and needlessly.

Among the early themes that have shaped and limited care are the following four myths:

Myth #1: Lyme is overdiagnosed. In the early 1990s, Alan Steere, the physician credited with discovering Lyme disease and an author of the first treatment guidelines, studied the records of nearly eight hundred Lyme-diagnosed patients. Fully 57 percent did not suffer from the malady, he found. As significant, nearly half of patients had tested positive in other labs and negative in Steere's. While such findings could have been taken as a tip-off of the poor state of Lyme diagnostics, they were read as a warning sign of overtreatment, one that would produce a stream of easily replicated science.

The Steere study, published in 1993, was followed by at least ten other published articles on overdiagnosis through the rest of the 1990s. Each came to similar conclusions and, in turn, led to more studies. Overdiagnosis leads to "frequent minor adverse drug events," said a study in *Annals of Internal Medicine*, such as nausea, rash, and vaginal itching. Ordering unneeded tests, it asserted, also eats up "considerable health resources," a sum that was later put at $492 million for more than three million tests every year. By 2016, about thirty publications in the National Institutes of Health library directly or indirectly addressed Lyme disease overdiagnosis. These included several studies of purported overdiagnosis in children and in people with neurological involvement, the kind of cases careful doctors might not want to miss. Just five studies addressed potential underdiagnosis.

On the overdiagnosed side were patients who falsely tested positive—there were no doubt some—and were prescribed antibiotics for short periods. On the underdiagnosed side were patients who had falsely tested negative for a disease that, once missed, can cause lasting pain and debilitation. Science sided with the folks who should be spared unnecessary treatment, which is logical. First, do no harm. The assumptions of those studies are open to question, however. First, they rested on the premise that the risk and cost of treating appreciably outweighed not treating, a question no study asked. Second, the studies presuppose that science can prove, through standard blood sampling, who actually has Lyme disease. That brings me to the second myth.

Myth #2: Lyme disease testing is reliable. The CDC endorses what is called "two-tiered" testing, saying on its website: "Patients who have had Lyme disease for longer than 4–6 weeks, especially those with later stages of illness involving the brain or the joints, will almost always test positive." Physicians have taken pronouncements like that, combined with prolific warnings of overdiagnosis, to mean one thing: Go by the tests. Yet dozens of scientific papers paint the Lyme test, which I'll discuss in chapter 6, as anything but cut and dry. Because of the time needed to muster antibodies, just 30 to 40 percent of infected patients will test positive early in the disease, a Lyme fact acknowledged by the CDC and IDSA.

Testing accuracy improves later on, but only after patients are sicker, the disease more advanced. And still cases are missed. Perhaps one in four patients with disseminated disease and one in eight with neurological disease will wrongly test negative weeks to months into the disease, according to key testing studies. Then there is the question of whether published accuracy rates, which vary widely, are even reliable. Studies that have validated the tests suffer significant biases, for example proving the technology works by testing it on people who previously tested positive. What about those it failed? Wide variation in accuracy has also been found depending on which commercial test, and in what two-tiered combination, is used. One may ring positive; another negative. The only definitive symptom on which doctors are encouraged to confirm a Lyme diagnosis is the erythema migrans, or EM, rash, and it appeared, according to a CDC study of 150,000 cases, in 69.2 percent of patients. That's three in ten cases potentially missed early on in a diagnostic milieu rife with testing pitfalls.

In 2016, the journal *BMC Infectious Diseases* published a review by a team of twenty-one European researchers of seventy-eight studies of the standard tests to diagnose Lyme disease. It was the most exhaustive review of Lyme testing ever done, analyzing studies of technologies commonly employed in the United States. Ultimately, the analysis found many test studies wanting, and all test studies plagued by at least one serious bias. The team's article concluded, "The data in this review do not provide sufficient evidence to make inferences about the value of the tests for clinical practice." In other words, six and a half dozen studies could not settle the issue: We don't know for sure if the tests work in a patient-physician setting. How can we say Lyme is overdiagnosed when we cannot trust the tests?

Myth #3: Lyme disease is hard to get. In June of 2001, a study was released that prompted the *New York Times* to publish a page one story that said, in essence, the nation's biggest vector-borne disease was nothing to worry about. It began: "Lyme disease is very difficult to catch, even from a deer tick in a Lyme-infested area, and can easily be stopped in its tracks with a single dose of an antibiotic, a new study shows." The study, published in the *New England Journal of Medicine* (*NEJM*), tracked 482 people in Westchester County, New York, who were given a prophylactic dose of doxycycline after they had been bitten by ticks. Just 0.4 percent of the group that received the dose developed Lyme disease. At the same time, 3.2 percent of a control group developed a rash symptomatic of the disease. The study made two points: Lyme disease could be prevented after a tick bite. And a small number of bitten people, three in a hundred, would get Lyme disease.

There is little dispute that a smaller share of infections result among people who know they have been bitten; ticks are more likely to be removed before they infect. Beyond this, the finding that a single doxycycline dose wards off disease came with a significant caveat. Researchers used the rash alone as proof of infection, admitting that doing so "could have resulted in underestimation of the actual incidence" of the disease. Indeed, some 30 percent of Lyme patients in the CDC study did not get the rash; at least one nonrash participant in the treated group tested positive but was not added to the single treated patient who nonetheless got Lyme disease. Nor were the study's patients followed beyond six weeks, when others might also have tested positive. Last, fifty-one people, 11 percent, dropped out of the study but were assumed not to have gotten the rash. A critique published in the journal *Expert Review of Anti-Infective Therapy* argued that just one more rash in the treated group would have erased the study's significance. The critique was included in treatment guidelines of the International Lyme and Associated Diseases Society, which favors twenty days of prophylactic treatment for tick bites based on studies that came to other conclusions. In 2004 and 2008, the single dose failed up to half the time to prevent Lyme disease in two mouse studies, while a longer-acting antibiotic injection was completely effective.

Despite the *NEJM* study's limits, its single-dose prophylactic finding went on to become the standard of care for tick bites. The tough-to-get, easy-to-stop message should help allay "inflated public fear of Lyme disease," the *Times* reported, while the *NEJM* editor, Dr. Jeffrey M. Drazen, pronounced, "This is reassuring information for people who make decisions based on evidence."

The study's release, along with the results of two treatment trials, was an early and clear attempt to tamp down a growing Lyme-patient insurgency, and it would not be the last. Beyond this, it was science put to the task of public relations. Its message was control—of an epidemic, with a questionable prophylactic, and of an alarmed public, with a single, oversold study that has yet to be replicated.

In the ensuing years, the Lyme disease toll would grow in the United States from about 17,000 in 2001 to 38,000 in 2015. But because of underreporting, those official figures represented just a tenth of all cases, the CDC announced in 2013. That year, Lyme disease was ranked fifth in the United States among reportable diseases even before the tenfold adjustment, especially impressive considering that 95 percent of reported cases come from just fourteen states. Do these numbers fit the description of a disease that is hard to get? Nonetheless, a physician quoted in the article, Dr. Leonard H. Sigal, who would go on to publish widely on Lyme disease overdiagnosis, said catching the disease wasn't the problem. "The bigger epidemic," he said, "is Lyme anxiety."

Myth #4: Lyme disease is easy to treat—and is not a chronic disease. "Patients treated with appropriate antibiotics in the early stages of Lyme disease usually recover rapidly and completely," says the CDC, reflecting the IDSA guidelines for care. Whether the infection survives antibiotic treatment to become chronic is the single most contentious issue in Lyme disease, one that the CDC and IDSA have long sought, unsuccessfully, to put to rest. When the Westchester County study on prophylactic treatment was released, the *NEJM* announced a second article on twin treatment trials, articles that were "so important," the *Times* reported, that they were publicly unveiled a month ahead of publication to coincide with the onset of summer and outdoor activities. In both trials, researchers gave ninety days of intravenous and oral antibiotics to Lyme disease patients who, although previously treated, had suffered "considerable impairment of health-related quality of life." The researchers ended the trials early, however, concluding that the treated group did no better than a group that got placebo drugs. "Prolonged and intensive treatment with antibiotics, a course of care advocated by a small group of doctors, does nothing for people with symptoms often attributed to chronic Lyme disease," the *Times* reported.

Although those two trials, led by Mark Klempner of Boston University School of Medicine, were used to assert that continued antibiotic treatment is unsuccessful and dangerous, the studies were the antithesis of a slam-dunk for Lyme practitioners. Treating advanced Lyme disease had long been more art than science, as these doctors built on clinical experience and shared knowledge to treat patients no one else would. One antibiotic combination might work for patient A, another combination of drugs may help patient B. Lyme practitioners also knew that tick-borne coinfections, which the trials did not consider, often immensely complicated diagnosis and care. They saw patients improve.

In 2012, a team led by Allison DeLong, a Brown University statistical researcher, analyzed the methodology and findings of Klempner's two treatment trials and two others by Lauren Krupp of Stony Brook University in 2003 and Brian Fallon of Columbia University in 2008. Like Klempner, Krupp and Fallon had reported mixed or fleeting patient improvement from additional rounds of antibiotics in studies that largely discounted their effectiveness. The review, published in *Contemporary Clinical Trials*, found otherwise. One of Klempner's two trials used "unrealistic" measures, DeLong found, deeming patients to have improved only when they surpassed the norm for the US population by essentially one standard deviation (placing them in the 85th percentile for health). Sample sizes were also likely too small to detect meaningful improvement, the analysis asserted. In fact, all four studies suffered statistical and analytical flaws that underestimated their positive effects on patients—in particular, in Krupp and Fallon, respectively, on fatigue and cognitive functioning. "Our careful examination of the trials suggests that, for some patients with Lyme disease, retreatment can, in fact, be beneficial," the review concluded. Fallon himself revisited his and Krupp's trial results in 2012, concluding that demonstrable benefits had been overshadowed by side effects that perhaps could be ameliorated.

Cheering on Failure

What was striking when the Klempner study was released was the enthusiasm with which researchers touted trials in which no solution was found, in which their efforts to heal had failed. "If this had been a study looking at tuberculosis," Tom Grier, a Lyme disease patient, wrote in 2001, "would the study be stopped early and then be rushed to press to tell us that the patients did not feel any better after three months of antibiotics?" This is the reality of Lyme disease treatment research, then and in the nearly two decades since Klempner's research. What little treatment research there is has been used to reinforce the dogma of the one side that claims to know what works (short-course antibiotics) and what doesn't (anything longer). Researchers whose findings challenge prevailing science, as I'll discuss here and in other chapters, are marginalized to small journals, they have told me, or made to revise their conclusions. Some do not get published at all when the results suggest that Lyme disease may be bigger, more widespread, and more persistent than the prevailing side believes. These researchers struggle to get funding for research from government agencies whose review boards are dominated by scientists with accepted and traditional views of Lyme disease.

The United States may not export much in the twenty-first century, but its made-in-America vision of Lyme disease sells well in many countries around the world. Lyme is exaggerated. Lyme can be dealt with. Lyme is no big deal. That has been the theme of paper after paper, in particular in the *New England Journal of Medicine*, where Klempner, the lead author on the treatment studies, has served as an associate editor for ten years. In 2007, the journal ran an "appraisal" of chronic Lyme disease, in which it reviewed the evidence against the use of antibiotics for lingering symptoms of Lyme disease, repeating, as other articles and commentaries have, the primacy of the Klempner findings. The timing of the review's publication was interesting, arriving after Connecticut Attorney General Richard Blumenthal had opened an antitrust investigation of the Infectious Diseases Society in November of 2006. The probe centered on potential conflicts of interest involving the physicians and researchers who wrote the guidelines that dictate Lyme disease care. Seven months after the appraisal was published, Blumenthal, who went on to become a US senator, delivered this stunning conclusion: "The IDSA's 2006 Lyme disease guideline panel undercut its credibility by allowing individuals with financial interests—in drug companies, Lyme disease diagnostic tests, patents and consulting arrangements with insurance companies—to exclude divergent medical evidence and opinion."

The *New England Journal* article was seen as an attempt to ward off the Blumenthal probe, with the nation's foremost medical journal employed in the service of pushback and public relations. The tone of the appraisal was indignant and some of the content had little to do with science. "Physicians and laypeople who believe in the existence of chronic Lyme disease have formed societies, created charitable foundations, started numerous support groups (even in locations in which *B. burgdorferi* infection is not endemic), and developed their own management guidelines," the appraisal stated. "Scientists who challenge the notion of chronic Lyme disease have been criticized severely."

Yet such criticism was nothing compared to what doctors who practiced outside the guidelines faced. At the time, several Lyme physicians in New York State—including those who had formed their own medical society and developed their own guidelines—were being investigated by the state's physician licensing board for prescribing antibiotics outside of the IDSA's treatment guidelines. These doctors told me the reviews had often lasted for years and cost them huge sums in legal fees. One physician had been suspended and another was cited for lapses in record keeping, insignificant infractions that were enough to scare other doctors away from treating difficult cases of Lyme disease. Similar actions were playing out in other states and a few countries, too. To guideline adherents, this rogue behavior was simply not acceptable. "Chronic Lyme disease, which is equated with chronic *B. burgdorferi* infection," the appraisal concluded, "is a misnomer, and the use of prolonged, dangerous, and expensive antibiotic treatments for it is not warranted."

The appraisal's authors included many original guideline writers as well as a CDC microbiologist, Barbara J.B. Johnson, who had long published with IDSA Lyme researchers and represented their viewpoints within the agency. Another lead author was Susan O'Connell, a microbiologist who headed the British health services' Lyme Borreliosis Unit, assuring guideline writers that their dictates concerning Lyme disease would be upheld in the United Kingdom as well. Similarly, scientists from Slovenia, Sweden, Denmark, and Austria have at various times published with the authors of the United States guidelines for Lyme disease care, facilitating the export of a prime American product: medical knowledge endorsed and promoted by the highest echelons of US scientific publishing and government.

So close are these ties that in 2016 the *New England Journal of Medicine* published the results of yet another treatment study, conducted in the Netherlands, which debunked long-term treatment for Lyme disease. The study used a similar treatment regimen of previous, supposedly failed, trials even as emerging science was questioning the effectiveness of such drugs. Moreover, the study acknowledged uncertainties and mixed findings: Patients in all three treatment groups showed early improvement, while, at fourteen weeks posttreatment, researchers saw significant physical improvement but not in quality of life. The journal's brand overshadowed these questions, as it long had. The overall findings that more antibiotics "did not have additional beneficial effects on health-related quality of life" were hailed in an *NEJM* editorial that called the study "well-performed," while its findings were widely reported as sealing the deal against longer treatment.

In November 2011, Britain's General Medical Council ruled that a general practitioner from the north of the country near Manchester had "diagnosed an infection similar to Lyme disease in patients with chronic fatigue syndrome (and) adopted an 'unwavering mindset' that ignored mainstream medicine." For prescribing long-term antibiotics and other unconventional treatments, Dr. Andrew Wright, one of the area's only physicians to treat advanced Lyme disease, was found to be "impaired" to practice. Around that table in the Winchester pub, UK Lyme patients told me of being cut off from antibiotics that had helped them, seeing a succession of doctors unwilling to treat, and being referred for psychiatric care. One had tested negative long after he had had the Lyme disease rash, when the CDC maintains tests "almost always" turn positive. This single "impaired" physician had helped them. "I count my blessings that I found him," a woman with longstanding Lyme commented online. Wright made the mistake of using unsanctioned diagnostic tests and a procedure not approved for office use that carried risk. But his real offense, like the New York doctor cited for record-keeping errors, was failing to follow the Lyme treatment guidelines. A French physician likened the atmosphere surrounding this disease to the Inquisition. Indeed, as I read the statement on Dr. Wright, in the context of diagnostic uncertainties and evidence of *Borrelia burgdorferi*'s considerable complexities, it reminded me of historical artifacts from an era of heretics and witches. Soon, I believe, it will be viewed that way.

Monkey Business

Starting in the first decade of the twenty-first century, the other side of the Lyme divide, in that parallel universe that had long agitated on the edges of the debate, began to gain credence and produce research breakthroughs. One of these came from a young scientist at Tulane University, Monica Embers, who had used *Borrelia burgdorferi*-laden syringes to infect five Rhesus macaques, which are small brown monkeys known for their pink, expressive faces and big eyes. Months after treating three of the animals with oral antibiotics, Embers and her colleagues then looked for evidence that the infection had survived. Among their methods, they used a technique called xenodiagnosis, in which uninfected ticks were put onto the treated monkeys and allowed to feed. Then the ticks were dissected, their mid-gut contents subjected to a stain and put under fluorescent lights. Sure enough, samples from ticks that fed on two of the treated monkeys lit up.

There, on a midnight black background in living lime green appeared more than a few coiled, curved, and entirely whole spirochetes, organisms that were supposed to have been dispatched by modern-day drugs.

Evidence of *Borrelia* DNA also was detected in the monkeys themselves, reinforcing findings of persisting post-treatment infection. True, the animals had been artificially infected, rather than sickened by tick-bite. "Our studies do however offer proof of the principle that intact spirochetes can persist in an incidental host comparable to humans, following antibiotic therapy," wrote Embers in *PLOS One*, an online open-access journal. Significantly, the experiment had used ninety-day antibiotic treatments, about three times longer than the guidelines recommend.

Several years earlier, Mario Philipp, who initiated the Tulane study, had infected twenty-four other macaques but was less successful at finding evidence of persistence in the twelve that were treated; he did not use the technique that Embers did. But he did find something else among animals given injections chock full of Lyme spirochetes. Seven of twelve monkeys tested negative 63 to 94 weeks later without ever being treated, lending credence to assertions that patients often do not test positive for infection when, in fact, they are.

Embers et al. weren't the first to show that the Lyme spirochete survived antibiotic treatment. In the early 1990s, a European study found live *B. burgdorferi* in the skin of five of twenty-eight people three months after they were treated with antibiotics; at least one other European study isolated the organism from a single treated patient. In animal studies over the previous two decades, Lyme spirochetes had been found to withstand antibiotic onslaught in dogs, mice, and other primates. Those animal studies generally based their conclusions on finding DNA of the spirochete, which is an accepted diagnostic tool, rather than by finding the bug itself. This time, however, Embers recovered actual spirochetes. Hers and the emerging research of others—by scientists at Johns Hopkins, Columbia University, UC Davis, and Northeastern—strengthened the argument that Lyme disease wasn't always cured by the limited antibiotic courses vigorously defended by guideline writers. Four years later, a Tufts researcher, Linden Hu, would use xenodiagnosis to recover

B. burgdorferi DNA from ticks attached to two of twenty-three Lyme patients; the only place they could have picked it up was from people who were supposed to have been cured. Nonetheless, Hu's team concluded, cautiously, that it did not mean "viable spirochetes" were in the patients themselves.

At a conference in New York City in 2016, I met a physician, Nevena Zubcevic, who had completed her residency at Harvard Medical School's Spaulding Rehabilitation Hospital in Boston. There, where many of the Boston Marathon bombing victims had been treated in 2013, Zubcevic saw patients come in with all the signs of having suffered traumatic brain injuries—neurological deficits, functioning problems, pain. To her surprise, she concluded many were actually suffering from Lyme disease. Moreover, these patients told stories of having been turned away by many doctors who, following standard protocols, had been told they should not give additional antibiotics so gave nothing at all. "They couldn't go on," Zubcevic told a group at Martha's Vineyard Hospital in 2016, on an island off Massachusetts where Lyme disease rates are among the highest in the United States. "Some patients show symptoms of posttraumatic stress disorder because they've been ignored for so long. Marriages dissolve all the time because one spouse thinks the other is being lazy. Many chronically ill patients end up alone."

In 2015, Zubcevic helped found the Dean Center for Tick-Borne Illness, based at Spaulding. Several decades into the epidemic of Lyme disease, it was the nation's first, and still only, rehabilitation center for Lyme disease patients. This is startling when one considers that well over two million people have been infected in the United States just since 2004, with perhaps 200,000 to 400,000 suffering lingering post-treatment problems.

To Zubcevic, Lyme disease care is complicated by poor diagnostic tests, multiple species of *B. burgdorferi* that manifest with different symptoms, and other pathogens that are delivered in a single tick bite. These include coinfections from *Babesia* and *Bartonella*, which the Klempner study never addressed. "This is very complex and messy, and I work with that mess every day," she told the conference. "I don't have the privilege of just working with Lyme disease. I wish that I did. My job would be a lot simpler." When I asked whether she followed the prevailing treatment guidelines, Zubcevic was firm: "I don't practice cookie-cutter medicine," she said. "We have to go beyond these guidelines."

"The Lyme Loonies"

In my research for this book, I was always quick to say, since I was often asked, that I was not a long-term or chronic Lyme sufferer. A hastily treated Lyme infection occurred after I began researching the disease in 2012, while another had been resolved years earlier. Both were signaled by rashes. In the controversy over Lyme disease, those seen as having vested interests—mostly the sick and those willing to treat them—have long been denigrated and dismissed. In 2013, I obtained several thousand pages of Centers for Disease Control emails that revealed a disdain for, even a fear of, Lyme patients. In one, an official at the National Institutes of Health wrote to a CDC microbiologist in 2007 to say goodbye upon his retirement. "I will certainly miss all of you people—the scientists," wrote Phillip Baker, the NIH Lyme program officer, "but not the Lyme loonies." In another email, an author of the guidelines commented to CDC and IDSA colleagues on a public protest by Lyme patients in Connecticut. "We need to mount a socio-political offensive; but we are out-numbered and out-gunned," wrote Durland Fish, a Yale University entomologist. In a separate email, Gary Wormser, the lead guidelines author, instructed his CDC and IDSA coauthors on an upcoming article to emphasize the risks of long-term antibiotics: "superbugs, death from treatment, biliary disease, cost, etc etc would be good," he wrote. There was very little room for opposing views in the echo chamber that is the

CDC–IDSA relationship. Patients who had a different view of Lyme disease were a bother there, and so were their doctors.

Testament to the parallel universes I have noted, I did not see a single major Lyme disease researcher from the mainstream side at the Mount Sinai conference at which Dr. Zubcevic spoke. The gathering was not supported by the CDC or National Institutes of Health but by a nonprofit group, the Steven & Alexandra Cohen Foundation, which had also underwritten many of the researchers who presented. These were scientists who believed in something other than the traditional view of Lyme disease and who had plowed ahead largely without government support.

Ying Zhang, a Johns Hopkins University researcher, was one of them. Zhang had done pioneering work in the early 1990s on the mechanism that allowed tuberculosis to survive antibiotic treatment, which represented a leap forward in TB understanding. Starting in 2014, Zhang and his research group published seminal articles on the failure of frontline antibiotics—doxycycline, amoxicillin, ceftriaxone—to kill what have been called Lyme disease "persisters." This and other test-tube research offered a compelling explanation for why patients remained sick. The organisms changed shape, hid, or otherwise went dormant—the persister cells—in their quest to evade elimination.

Yet Zhang, a scientist with an international reputation at a top university, was rejected for government grants and publications. Reviewers, he told me, said it wasn't clear if "the few persister organisms" found in such studies caused ongoing symptoms or if patients even had "bona fide Lyme disease" in the first place. Similarly, Richard Ostfeld, a leading researcher on the interplay between ticks and the hosts that support them, was denied funding in a review that called Lyme a "middle-class disease." So was Monica Embers, the monkey researcher, whose grant reviewer, she recalled, commented, "Lyme isn't that much of a problem, and there's a vaccine anyway," which there wasn't. Indeed, Embers' paper on chronic infection in Rhesus macaques was rejected by at least three journals before finding a home in *PLOS One.* "I absolutely expected it because I knew we were challenging dogma in a way," Embers said in relating the story. That dogma still held sway even though, as far back as 1997, a young doctoral student at Cornell University, Reinhard Straubinger, had reported the spirochete's survival in dogs treated with antibiotics, and researchers since had found spirochetes in treated gerbils, rats, hamsters, and guinea pigs. Stephen Barthold, a scientist at the University of California, Davis, and member of the prestigious National Academy of Medicine, had observed the phenomenon in mice in 2008, yet he also faced what he called "prejudicial" reviews for grants.

European scientists have also faced such hurdles. "I've tried to publish in many journals," Christian Perronne, a physician on the infectious diseases faculty at the University of Versailles Saint-Quentin, France, told a conference in Norway in 2014. "If you try to publish a little bit different from the guidelines, it's anti-science." He told the story of a colleague who had lost a university position after reporting substantial rates of infected ticks in the countryside. "Maybe I will be next," said Perronne dispassionately. This from a physician who had served as president of the Communicable Diseases Commission at the French High Council for Public Health and vice president of a working group on European vaccination policy for the World Health Organization.

Yet his and other science, which has time and again bucked the traditional view of Lyme disease, has had to claw and scrape to break through amid denial and dismissiveness. Even letters that have questioned conventional published research, scientists have told me, are routinely rejected by journal editors. The experiences of these and other scientists, who said they approached their Lyme research with healthy skepticism, are emblematic of the ways the disease has been pigeonholed and minimized. They are rooted in the "easy-to-treat" mentality, the group-think that, while acknowledging that some patients suffer a vague array of problems afterward, sees antibiotics as curative and Lyme disease as a single, surmountable disease.

As Zubcevic's patients attest, the disease is anything but. Kim Lewis, a distinguished professor and director Northeastern University's Antimicrobial Discovery Center, found evidence in laboratory experiments published in 2015 that the Lyme spirochete had survived recommended antibiotics like doxycycline. A Moscow University trained microbiologist, Lewis spoke at the Mount Sinai conference of Lyme disease patients who do not get well after treatment. "Interestingly, when you show up at the doctor's office," he said, "nobody tells you that you have a 10 percent probability that something very unpleasant is going to happen to you, and maybe we won't have any remedy for that. We'll just give you some doxycycline and you'll be fine. A lot of people are not."

What Embers, Perronne, Zhang, Zubcevic, and Lewis were fighting was the image of Lyme disease enshrined in the IDSA treatment guidelines—and American medicine. The guideline writers, a close-knit, like-minded group, did solid work on the early phases of the disease. But their vision was stuck there while science evolved, their guidelines rife with their own "expert opinion," which is the basis of half their recommendations. Following suit were the agencies that should have been funding research on the later and protracted manifestations of Lyme disease. "The early part of the disease is only a part of the story," Zhang told me. "The IDSA line is very restrictive." So, too, is the official view of Borrelia burgdorferi's human toll, as captured in the CDC's proclamation that only a "small percentage" will have ongoing problems after treatment. Nothing to fret about, unless you are in that group, which may not be so small.

In 2005, a statistician from Germany and an epidemiological researcher at Oxford University analyzed five published studies on the after-effects of Lyme disease. Comparing 500 patients to 500 controls, the researchers found that rates were "significantly higher" among Lyme disease patients for "fatigue, musculoskeletal pain, and neurocognitive difficulties that may last for years despite antibiotic treatment." In 2015, four researchers at the University of Freiburg in Germany, reviewed thirty-four studies on post-Lyme conditions for an article in the *Journal of Neurology*. They found that 28 percent of patients suffered continuing symptoms after treatment for Lyme neuroborreliosis, a form of the disease in which the spirochete has infected the nervous system, sometimes leading to severe headache, eye problems, lapses of memory and thought, depression, and facial palsy. Even the study used by the CDC to conclude the post-Lyme toll was small dutifully reported that eight other studies had found posttreatment rates as high as 41 percent. Moreover, the study's supporting editorial—the one that equated post-Lyme symptoms to Americanitis—included this caveat: The study "did not specifically include patients with…neuroborreliosis or late Lyme arthritis. Such patients could answer survey questions differently in the long term," it said. So why did the CDC rely on the under-5 percent rate—or on this paper at all?

Instead, in adopting the position that a "small percentage" of patients may experience posttreatment symptoms, the CDC brushed aside dozens of other studies that showed quite the opposite: Lyme lingered.

CHAPTER 5:

"Little Armored Tanks"

Bob Maurais runs a tick control company in Portland, a harbor city along the green coast of Maine, where he grew up in the 1950s and 1960s. Back then, Maurais spent a lot of time in the state's southern hardwood forests, trolling for natural treasure with three older brothers. He cannot recall a single blacklegged tick, or any other tick for that matter, from when he was a boy. Today, they thrive. In the winter of 2013–14, Portland received eighty inches of snow, had ten days of temperatures below zero, and the coldest March in fifty years. The tick population, responsible for a sixfold increase in Lyme disease over the previous decade, shook off its slumber, nonetheless robust and unscathed. "They are not going away anytime soon," he told me.

The alacrity with which Maine's ticks awoke plays out billions of times in forests and fields when winter turns to spring in the Lebanon Valley of eastern Pennsylvania, along Holland's North Sea dunes, and in bits of woodland from suburban Beijing and the forests of Poland to urban Chicago, London, and Washington, DC. On the scale of human time, the emergence of ticks has occurred at breathtaking speed, fostered in no small part by rising average temperatures. In the Adirondack Mountains of New York, as elsewhere, ticks wake increasingly at odd times of year: after a hard frost in December when daytime temperatures run unseasonably warm, in January amid a blast of balmy air, in an increasingly snowless February when ticks might ordinarily have been locked under a blanket of white.

I see a lot less snow in the Hudson Valley than I did when I attended college along the river in the early 1970s and then moved to the valley from New York City in 1982. Since 1970, annual average temperatures across New York State have gone up 2.4 degrees Fahrenheit in two generations. Winter warming has increased 4.4 degrees Fahrenheit, a stunning change in meteorological terms. In many parts of New York State, trees leaf out eight days sooner and buds bloom four days earlier than in the 1950s. Bees arrive to pollinate ten days earlier than they did in the 1880s. Breeding birds in the state have shifted northward, as have ocean fish along the state's coastline.

Ticks have long been hearty survivors of trial and calamity. A quarter century after the Chernobyl nuclear power plant meltdown in 1986, former Soviet-bloc researchers found *Dermacentor reticulatus*, members of the *Ixodidae* family of hard ticks, thriving within the plant's exclusion zone. In perhaps the quintessential example of a species' adaptability, the ticks, the researchers concluded, "can be maintained in areas after a nuclear disaster with radioactive contamination." A quarter of them, incidentally, were infected with the bug that causes anaplasmosis, a few with the one responsible for babesiosis.

From 1991 to 2012, researchers at the Cary Institute of Ecosystem Studies in Millbrook, New York, captured 54,000 mice and 12,100 chipmunks; they counted 403,000 larval ticks and 44,000 nymphs on the ears of those animals. In the space of a human generation, they found unambiguous changes in the ecosystem in which these creatures lived. As the climate warmed, the cycle of life for ticks was reshaped, their birthing and development occurring earlier with the passing years: "Accelerated phenology," the researchers called it.

Extended springs and summers mean that ticks arrive earlier in the year, of course. But the implications are greater still, affecting the interplay between tick stages of development. Larvae hatch and feed in late summer, when some will pick up infections if their host, usually a mouse, is infected. They then molt into nymphs and, if infected, go on to infect other animals, or people, when they feed the following year. Earlier springs means infected nymphs have a longer time to feed, giving more of them the chance to infect more animals, to spread and increase each animal's cargo of *Borrelia burgdorferi,* which will later be tapped by larvae. More infected animals means more Lyme-infected ticks and an ever-more perfect cycle to sustain an epidemic.

So compelling were the changes in climate seen in this one corner of the world that nymphs and larvae were predicted to hit their peak eight days to two weeks sooner by midcentury, the risk of tick exposure, the Cary researchers concluded, coming "substantially earlier." The study's final year, it turned out, was also its warmest. If trends continued, temperatures seen in 2012, the researchers wrote, will be "substantially cooler than normal by the 2050s." Yes, the hottest year in the study period would be cool by future standards.

A Dutch scientist I had visited in 2016 foresaw a time when ticks were a year-round phenomenon in a country once renowned for its frozen canals. Willem Takken at Wageningen University had studied tsetse flies and malaria in Africa and had turned his attention in 2000 to the arachnids invading his country. Seven to 26 percent of nymph ticks—the ones likeliest to bite people—were infected with Lyme disease. These ticks, carting other pathogens as well, were common along woodland trails, in homeowner gardens, and in the brush along roadsides. In the rapidly warming Netherlands, Takken said, "The idea that ticks go away is no longer sustainable."

Armored and Impervious

Short of frying them in the hot sun, blacklegged ticks are very difficult to kill, which is why, forty years into an epidemic of tick-borne disease, ticks populate and proliferate. I once washed a nymph down the bathroom sink and returned to find it crawling up the white porcelain bowl; now I close the trap. Try crushing one with your fingernails. Swatting them is out of the question. Holly Ahern, a microbiology professor at a state college in the Adirondack foothills of New York, spent a semester's project figuring out how to get the guts out of *Ixodes scapularis* ticks to test them. Granted, she was new to the task. Her students pressed, banged, and mashed them with mortars and pestle, all to little effect. Finally, placing one under a microscope, Ahern cut it with a scalpel below the hypostome, which is a barbed spear near the mouth that the tick inserts at mealtime. The salivary glands, south of the hypostome, are where the payload is in ticks, where all manner of secrets lie, so the goal was to empty them. Then, she used a power drill with a coarse grinder attachment, which worked unless the critter slid sideways and twirled around like a carousel pony. The blacklegged tick and its many relatives have ingenious protective qualities, the upshot of millions of years of evolution. Among these is a kind of shield on its back, sclerotized in entomological parlance, called a scutum for females, conscutum for males. These plates, and more on the tick's underside, are the very definition of hard-bodied.

Felicia Keesing is a research scientist from Bard College in Annandale, New York, who has studied *Ixodes scapularis* for years. She described the ticks to me as "tiny little armored tanks...exquisitely adapted to what they do," which, not incidentally, is "to drink your blood and infect you with parasites." For her, the problem isn't so much how to kill them, but how to find them. The ticks feed for just a couple of weeks in their entire two-, sometimes three-year lives, attaching to a host once for each life stage. They can go hundreds of days without a meal. We see them only when they quest for food—climbing up wildflowers, perching on low branches—and when they find a host and attach. "But we don't know where they are most of the time—the entire part of their life cycles when they're not on hosts," she told me. "And we're not good at getting to them during these long phases." Among *Ixodes'* many attributes, armor notwithstanding, stealth may be its biggest protection.

When ticks make an appearance, it is for a performance only nature could orchestrate. Thanks to the magic of cinematography and microscopy, researchers at Harvard and Charité University of Medicine, Berlin, recorded an *Ixodes ricinus* tick, the kind seen in Europe, as it pierced the hairless ear of a mouse. The drama might be called exquisite—and entomologists will shudder at this—if it served some useful purpose. *Ixodes* ticks appear to play no essential role in nature, as least not one that's known, as I will discuss later. On screen, the tick is seen using spear-like appendages to pierce its victim, extending one of two sharp tips to almost twice its usual length, retracting it, and then extending the other. As the cameras whir, the tick rhythmically alternates this poking and prodding, until, voilà, the epidermis is breached. A toehold is planted. Then the twin spears, called chelicerae, snap into full breaststroke mode, flexing and retracting, to plant the tick's barbed hypostome—its anchor—on mammalian turf. The researchers demonstrated what to that point had been theorized: "The cheliceral teeth are thought to be adapted to cutting as well as holding, and ticks are said to 'cut,' 'saw,' or 'push' their way into the skin of their hosts," they wrote of their adventure in cinéma vérité. Yes, indeed.

Ticks, like mosquitoes, are small. But many people who are bitten by ticks, unlike those jabbed by mosquitoes, report that they never felt the bite, never knew when that offending arachnid geared up its chelicerae, inserted its hypostome, and began to sup. There is a reason for this, another wonder of nature that ought, in my mind anyway, to find a better use. *Ixodes* tick saliva is a feat in itself, laced as it is with an anesthetic, a Novocain to numb the skin, which makes the host oblivious and neatly heads off any attempt to swipe the tick away. Other molecules in saliva prevent blood from clotting and ending an otherwise hearty meal, while still another salivary substance defuses efforts by the immune system to jettison the parasite. After it is firmly in place, the tick builds a hardened seal around the bite using—what else?—its saliva. Like superglue, this cement locks the tick firmly in place for feeding, usually for three to four days.

During that time, the adult female's reproductive organs and salivary glands develop, and her outer covering, or cuticle, grows with her girth. When she starts feeding, she is a taut, walnut brown freckle with an orange crescent rim, flat and indestructible. When finished, she has increased her body weight several hundredfold and morphed into a squishy grayish grape with coiled legs, fat and ready to drop off—sometimes with her tiny slim mate still attached—and lay some 2,000 eggs. Of note, the males get their cake but don't eat it at this stage of life. They may mate several times, apparently preferring a nicely rotund partner, but they feed little, if at all, as adults.

In the second week of September in 2016, I spent an afternoon with three Cary Institute wildlife biologists, trolling for ticks in the brush and beaten leaves of a housing development on the outskirts of Poughkeepsie, a few miles east of the Hudson River in upstate New York. Wearing white coveralls, the explorers carried white corduroy flags on long poles. Slowly, they ran each flag over small patches of earth, hoping in thirty seconds to entice an appearance by the hardest tick to see: the blacklegged larval tick. The larvae are the progeny of those hefty females, their six-legged babies that hatch in summer, ready for their first blood meal.

It was late in the season when we went hunting for larvae and most had fed already or died. During one early season sweep, I was told, more than 2,000 larval ticks had dotted a biologist's flag, an institute record. Several years earlier, a friend had snagged 500 ticks and days later found several attached to the side of her breast. The few ticks that we snagged that day were so small as to be invisible to my untrained eye. Not too long ago, it did not matter much if a larval tick escaped detection; baby ticks, after all, were considered clean and uninfected. They had not yet latched onto a mouse, chipmunk, or other infected animal and been inoculated with a pathogen.

But there is an emerging bug that has ratcheted up the potential risk of tick-borne disease. Called *Borrelia miyamotoi*, the spirochete can be passed transovarially, meaning from tick mothers to babies, some born ready to infect with their first meal. The pathogen has been found in ticks in Europe, North America, and Japan, where it was discovered in 1995, a concerning trend for other reasons. *B. miyamotoi* infection rates, for one, were twelve times higher in female ticks attached to deer than those not, a 2016 study found. This suggests that deer, which normally don't infect ticks with the Lyme pathogen, may be good instead at inoculating them with *B. miyamotoi*, tainting the eggs to be laid later. The first human cases of the disease were reported in North America only in 2013, but the bug may have sickened folks far longer. It looks a lot like Lyme disease but usually without a rash. Moreover, there is no good test for this infection. When *B. miyamotoi* was found as frequently as the Lyme pathogen in ticks around the San Francisco Bay in California, researchers suggested, logically, that previous cases may have been missed. I thought of all of this as I looked around at where we were trolling for larval ticks—in a back-yard rimmed with bushes and late season weeds, in which, tucked in two places, I saw balls left behind by children.

These one-acre housing tracts are the proving grounds of ticks, where they test their mettle and earn their stripes. Here, in habitat heaven, is a plethora of passing mammals on which to feed and become infected—chipmunks, shrews, and mice in the eastern United States, for example, squirrels in the American West, Switzerland, and the United Kingdom. Add to this tableau the deer that love nothing more than munching on impatiens, pansies, hosta, and azaleas, all mainstays of middle-class gardens. Ticks are picked up here, let off there, like some suburban transit authority for arachnids. The home-and-garden lifestyle has abetted Lyme and tick-borne disease by merging the animals and flora that support ticks with the people who become infected by them. From 1985 to 2005, the United States saw a 90 percent increase in homes built on one-half to one acre of land. By 2014, the median size of a subdivision was twenty-four acres and sixty homes, a suburban ideal that eats forests, hurts wildlife, and enables—and puts people in the path of—ticks.

Make no mistake, however, this is not only a middle-class disease, just as ticks are not limited to tract developments. Hikers in stands of California oak; forestry workers in Spain; garden laborers, mushroom pickers, children in day camps and on school playgrounds; urban park visitors; the folks who maintain highways and utility lines, who mow and plow and seed. The opportunities are all-too rich, in many parts of our modern, managed world, to come into contact with infected ticks.

Carved-Up Nature

In 2003, a World Health Organization report on climate change included a list of diseases with the key factors that were driving them. Malaria was fostered by deforestation; hantavirus by growing rainfall. Cholera was being fed by urban crowding, river blindness by dams and irrigation. Next to Lyme disease was a curious driving factor: reforestation. Not deforestation. Reforestation. This is the paradox of Lyme disease, and it has received far more attention than it deserves. It is true enough, but not the real story of Lyme disease. Planting trees surely helped bring back the deer that host adult ticks. But reforestation did not create an epidemic. The ecology of Lyme disease, as discussed in chapter 3, is much more complex than that.

In the last half of the nineteenth century, each and every day, 8,400 acres of forest in the United States were cut down, plowed over, or otherwise built on. The blood-letting leveled off in the twentieth century, and some of the nation's forest cover returned, with a net gain of about 12 million acres from 1910 to 2012. But the regrown forests in the US Northeast and Midwest, and in Europe as well, the woodlands that allowed deer once again to thrive, were not like the wilderness that preceded them. They were changed in quality, character, and size. They were, and are, subpar fragments of their richer, original whole.

The United States Forest Service analyzed forests in the contiguous forty-eight states from 2001 to 2006, and in a five-year period found a net loss of 1.2 percent of woodlands. More disturbing, however, was a 4.3 percent decline of what it called "interior" forest, wild land in parcels big enough, at least forty acres, to sustain processes essential for forest health. Pictures taken from satellites in space revealed ever more forested land broken up by parking lots and suburban tracts, in what the Forest Service researchers called "clustering patterns." In this version of modern day forests, species are lost, gene pools isolated, and habitats degraded. The loss of forest is one thing, the slicing and dicing, another. "The fragmentation of forest area into smaller pieces changes ecological processes and alters biological diversity," the Forest Service report said. And that allows ticks to thrive.

In 2000, a student at the University of Michigan in Ann Arbor named Brian Allan studied tick populations in different sized patches of forest, from about two to eighteen acres, in suburban upstate New York. The results were an indictment of the toll of middle-class development on human health. Taking a census of ticks, Allan found remarkable differences depending on woodlot size. Compared to bigger parcels, the smallest patches had three times the number of nymph ticks and seven times the number infected with the Lyme pathogen. Moreover, the share of infected ticks in the smallest patches was frighteningly high—seven in ten on average—which was the highest recorded to that point. The reason for this suburban curse goes back to what these islands of nature can support and what they cannot. Mice and other small mammals abound. Predators that might eat them do not. And this arrangement works very well for ticks.

In 2012, Taal Levi of the University of California, Santa Cruz, led a team that did two things to challenge the deer-foments-Lyme theory that has been linked to regrown forests. First, researchers analyzed twenty-one studies from four European countries and seven American states and found very mixed results in real-world tests. In some cases, deer culling led to tick reduction; sometimes it didn't. Sometimes, as in Finland, the Lyme bug was found in ticks where no deer lived. Deer enclosures in Lyme, Connecticut, meantime, greatly reduced young ticks but produced mixed results for adults.

Then, the team used a sophisticated statistical model that compared county-level Lyme disease cases, deer abundance, and populations of coyotes and foxes in four Midwestern and Northeastern states. The problem, researchers concluded, was not one of too many deer. There were, instead, too few red foxes, a primary predator of the four small mammals—white-footed mice, American chipmunks and two kinds of shrews—that infected 80 to 90 percent of ticks. In New York, Lyme disease cases rose along with increasing numbers of coyotes and declining numbers of foxes. Why? Coyotes eat foxes; foxes consume mice. On the deer question, meantime, the researcher found a link, though weak, between the size of the deer herd and the toll of Lyme disease cases in Virginia. But the clear opposite was noted in Wisconsin and Pennsylvania, where more cases sprang up in areas with fewer deer. The findings, published in the *Proceedings of the National Academy of Sciences* in 2012, aligned with Allan's research into tick-infested forest fragments. Many mice, few predators, and generally less diversity meant more ticks and more disease.

In the late 2010s, a pest-control company asked, in a tick-control advertisement, "Can you get Lyme disease from acorns?" In 2006, researchers at the Cary Institute reported this revelation: A forest flush with acorns one year led to many well-fed mice the next, and to many hungry ticks the following year. This, moreover, was no random act of nature but an upshot, more likely, of climate change. Warmer seasons and more carbon dioxide are kind to the production of tree seeds. This leads to more banner years for acorns, and with them, more mice and more disease-infected ticks. Deer? A simple answer to a complex question.

"Death Pursues the Aboriginal"

In the mid-2000s, a team of British and American experts in ecology, economics, conservation medicine, political science, and veterinary epidemiology was assembled to record, by time and place, the number of diseases that were emerging in rapid-fire fashion around the world. It was led by Kate E. Jones, an evolutionary biologist at University College London whose professional biography notes she is "allegedly*" eighth cousin to Charles Darwin, six times removed. The asterisked qualifier gives the source, as any self-respecting scientist might, as the UK website for ancestry.com. Jones had researched the fate of the world's 1,000 species of bats—no small feat—identifying the forces, such as loss of tropical forest, that were driving some to extinction. Now she wanted to take a similarly sweeping look at the scope of global disease outbreaks.

A half-century earlier, humankind had more or less declared a truce with infectious diseases, a hard-fought struggle aided by vaccines, antibiotics, pesticides, and public health monitoring. All of that changed by the late twentieth century, in what two scientists, British and Australian, have defined as the earth's "fourth major transition" in infectious disease. The first transition, wrote Robin A. Weiss and Anthony J. McMichael in the journal *Nature Medicine*, began some 10,000 years ago with the development of agriculture, domestication of animals, and growth of communal societies; that's when measles and smallpox emerged. The second transition, beginning in the centuries before and after Christ, was spurred as Eurasian societies came together for trade and war; rats with fleas and humans with lice carried typhus along. The third transition, around 1500, brought measles and smallpox to the New World. "Wherever the European has trod, death seems to pursue the aboriginal," Darwin wrote in "The Voyage of the Beagle" in 1839.

In the closing decades of the second millennia, a new age of disease was dawning, in line with other major and related changes in climate and lifestyle. Weiss and McMichael ticked off some of the reasons: urbanization, poverty, social upheaval, air travel, land clearance, and climate change. "There seems little doubt that more incidents are occurring, and have the potential to spread more widely than 50 years ago," they wrote. The scientists identified some of these new-era diseases: Legionnaires', HIV/AIDS, hepatitis C, bovine spongiform encephalopathy, severe acute respiratory syndrome (SARS), and avian influenza. They counted, by 2004, thirty emerging diseases.

As research by Jones' team would show just four years later, that figure was but a small sample, the leading edge of a bigger wave of illness. The tally was startling. From 1940 to 2004, the world experienced 335 emerging infectious disease "events," outbreaks that grew decade by decade in a bar chart that looked like a steep set of stairs. There were about twenty outbreaks in the 1940s, just over forty in the 1960s, and ninety-plus in the 1980s, a spike attributed to HIV-related diseases. By the 1990s, the tally approached 80. These disease events around the globe included outbreaks of new bugs that only recently had infected people like SARS; new strains of old pathogens that had become drug resistant, such as of malaria and tuberculosis; and, last, pathogens that had existed historically but soared in the latter twentieth century. In that final category, the researchers cited but one example: Lyme disease.

The study was the first scientific analysis to show that diseases were rising rapidly around the world. Beyond the sheer number, the array of epidemics hinted at the ecological and environmental forces behind them.

Sixty-percent of the diseases were zoonotic, meaning they pass between animals and humans; the researchers called this "a very significant threat to global health." About 23 percent of the diseases were vector-borne, namely illnesses transmitted by ticks, mosquitoes, and so on; that figure rose to 30 percent of emerging diseases in the final decade of the study, a rather unsettling trend. Echoing what other researchers have suggested about the rise of ticks, the Jones study pointed to climate change, which it identified as a potential factor in the proliferation of vectors, like ticks, that are "sensitive to changes in environmental conditions."

Damaged Assembly Line

In the early 1990s, Anthony J. McMichael, the Australian disease researcher, wrote a book on the threat to human health of overdevelopment, deforestation, biodiversity loss, and what he called "greenhouse warming." A major university publisher rejected "Planetary Overload," he told an interviewer in 2012, as "rather speculative and a bit fantasy-like...written from the vantage point of a privileged society member." By then, McMichael had been named to the National Academy of Sciences, and his views had been validated. "Climate change is not just about disruptions to the local economy or loss of jobs or loss of iconic species," he said two years before his death in 2014. "It's actually about weakening the foundations, the life support systems that we depend on as a human species."

As the Forest Service report put it, "It's possible that the conversion of forest land to development reduces both the product coming off the assembly line, and the assembly line itself." Cut-up chunks of forest, with their plethora of deer, mice and other mammals, and their paucity of natural predators, are part of that weaker foundation, that damaged assembly line.

And so, ticks march on. In 2014 and 2015, government scientists looked at nine national parks in the Northeast and Mid-Atlantic region, sampling for blacklegged ticks along well-used trails. In all nine parks, they found ticks infected with the Lyme pathogen. In addition, *B. miyamotoi* and another tick-borne bug that causes anaplasmosis were found in ticks in 60 and 70 percent of the parks, respectively. The national sites included Rock Creek Park in Washington, DC, the first time tick populations had been observed in the urban park. Here, five miles north of the White House, was where researchers found the highest density of infected nymphs, the ticks that most often infect people.

The findings published in 2017 were another indicator that ticks and the pathogens they carried had moved in all directions—north, west, and south—from the coastal Northeast where the disease first emerged. Several months earlier, Lyme-infected ticks were reported on the Outer Banks, a stunning crescent of tourist-trodden islands that hugs some 200 miles of northeastern North Carolina coastline. Beyond ticks, the bug was found in white-footed mice, rice rats, and marsh rabbits, a nicely constructed framework for pathogen production. While reported Lyme disease cases in the United States have largely been limited to states in the Northeast and Midwest, the Outer Banks researchers pointed to other evidence besides theirs of Lyme disease in the coastal south: infected ticks and rodents near Charleston, South Carolina; infected birds from islands off the coast of Georgia; infected ticks on barrier islands off the coast of South Carolina. "Lyme disease spirochaete transmission is occurring in a nearly continuous chain along the coasts of these three Mid-Atlantic States," wrote the researchers, led by Jay Levine of North Carolina University in Raleigh. Residents of those states, and in the South generally, are told that Lyme disease is rare, that it was likely contracted elsewhere, that the regional ecology isn't friendly to the support system needed to maintain *Borrelia burgdorferi* in the wild. Jay Levine's research suggested otherwise.

Ixodes ticks do not like hot sun so it's logical to assume they would more likely spread north, as the climate warms, than south. Testing that theory, innovative researchers from the United States Geological Survey and two universities mated Wisconsin males with South Carolina females, a tick marriage of Yankees and southern belles. Then they subjected the unwitting offspring to simulated conditions in the US South and North. Four-month-old larvae died faster under southern conditions. They concluded that a warming climate may limit *Ixodes'* damage in the South, keeping the ticks safely tucked in the leaf litter where they don't bite people.

That is not, however, what the southern state of Virginia is seeing. Holly Gaff is a biologist at Old Dominion University in Norfolk, which is on the Chesapeake Bay in Virginia's southern reaches, just thirty miles from the North Carolina border. When Gaff snags a handful of black-legged ticks on one of her white denim flags, this southern version of *Ixodes scapularis* usually curls up and falls off, sensitive as it is to heat. But farther north in Virginia, these ticks do something unheard of a decade ago. They crawl toward the human in reach. As heartier stock from the north moves into Virginia, she believes, "we are seeing displacement coming all the way down the state. Northern ticks outcompete these southern lazy guys." This may explain why Lyme cases rose from about 200 statewide in 2005 to nearly 1,300 in 2014. Indeed, cases have moved, presumably as infected ticks have, in a north-to-south migration, down the Shenandoah Valley and along the Blue Ridge Mountains. Gaff sees them slowly creeping around the Chesapeake to Norfolk, where she waits to snag them, corduroy flag in hand. The US Geological Survey paper that postulated ticks may stay north included a huge qualifier: it found more ticks surviving under southern conditions when humidity was high. In summer, Gaff said, humidity is always high.

Even if blacklegged *Ixodes* ticks limit their influence to cooler climes—a big unknown—the South is not, so to speak, out of the woods. There is a far more prevalent, and heartier, southern arachnid known as the lone star tick, long implicated as a vector for Rocky Mountain spotted fever and ehrlichiosis. As you will soon learn, this heat-loving, ubiquitous tick has been implicated in a similar disease in the South—Southern tick-associated rash illness—that in some, perhaps many, cases may in fact be Lyme disease.

Do They Serve a Purpose?

Research shows that eight-legged creatures—that cannot move more than a couple of feet, cannot see but sense, can go for months without eating, are hard to find and harder to kill, and carry diseases—may be coming to a neighborhood near you. Is there anything good about this, or more specifically, about ticks at all? Do they serve any essential biological purpose? I asked scientists who have studied ticks for many years.

Felicia Keesing, the Bard professor, recalled her studies in Kenya and her sightings of oxpeckers, small, dark birds with beige breasts, and depending on the type, red and yellow or all-red beaks, and rings around their eyes. The oxpecker would perch atop a large animal—rhinos, giraffes, bison, zebras—and scarf up the many engorged tick offerings of sub-Saharan Africa. The oxpecker burrowed into the ears, between the eyes, along the spine of the chosen animal. It was an exercise the beast tolerated in one of those cooperative ventures in nature, a kind of mutualism that safari-goers love. But do the oxpeckers need the ticks to live?

"What most people don't realize, though, is that the oxpeckers are really blood-eating creatures," Keesing said. "If the blood is inside of a tick, fine. But if there aren't blood-filled ticks around, the birds will gouge holes in the side of the wildlife and drink the blood directly." So maybe they do need ticks, and maybe they don't. As a 1999 article on oxpeckers in the journal *Behavioral Ecology* put it, "Their diet includes *Ixodid* ticks, dead skin, mucus, saliva, blood, sweat, and tears." They have eclectic taste, these birds, to be sure.

Closer to Keesing's New York home, no northeastern mammal or bird seems to depend on ticks as a dietary staple. Opossums will eat them strictly as part of a grooming regimen that helps control tick populations. Wild turkeys will eat them but not seek them out. Some new work has hinted that salamanders and wolf spiders might consume a lot of ticks. Who knows? "It might be that these species are real allies in reducing tick numbers," Keesing speculated. We can only hope—not that eating ticks is itself without risk. In Ireland, the population of red grouse, a squat brown bird with short tail feathers, has been decimated in recent years, possibly, researchers theorize, because they eat ticks infected with a virus that causes Louping Ill, a disease found in farm animals. In the process of grooming themselves, the grouse may be acquiring a disease with a mortality rate of 80 percent.

Maria Kazimirova, a Slovakian scientist who has studied the secrets of tick saliva, agrees ticks do not serve much of a role in the web of life. But perhaps we shouldn't be too quick to write ticks off, she suggests, at least in the pursuit of medical breakthroughs. Recall the ways that tick saliva numbs, suppresses its host's immune system, and has anticoagulant properties. These eight-legged menaces may have a future in pharmacological research.

Said Taal Levi, the UC Santa Cruz researcher, "I do not know of any purpose for *Ixodes scapularis* ticks." Indeed, ticks may not have a purpose from the human perspective, but they have, to be sure, a role. If not for the tick, *Borrelia burgdorferi* would be nothing. It would not get from here to there, would not be passed from mouse to deer and back, would lack the vector of vector-borne disease. We may not appreciate this as a magical thread in the web of life, but there it is. I won't be the one to write off one of God's less likeable creatures. Although I am surely tempted.

"Strictly from the point of view of humans," Kazimirova told me, "ticks cause more harm than benefit."

CHAPTER 6:
Faulty Tests

In 2001, *The New York Times Magazine* ran an article that set the tone for how Lyme disease patients, specifically those who disagreed with treatment policy and protocols, would be viewed for years to come. Entitled "Stalking Dr. Steere," the five-thousand-word feature began, "Last year, Dr. Allen Steere, one of the world's most renowned medical researchers and rheumatologists, began to fear patients." Back in 1977, Steere was the curious doctor, then thirty-three, whose investigations of a cluster of illness in Lyme, Connecticut, led to the discovery of Lyme disease. Now, the people he had tried to help were turning on him. He was a proponent of short-course antibiotics, and they blamed him when they remained sick and were not believed. It was a good story line, and the *Times* ran with it.

If it wasn't perfectly clear from the get-go, Steere was the hero of this story, a man with "a gentle, almost artistic temperament," and a "quiet and reserved nature." He was a "virtuoso violinist," "a kind of Magellan of medicine." The villains, mean and menacing, were Lyme disease patients. They stalked him in "hordes," carrying signs when he spoke in public that said terrible things, the story said, like "How many more

will you kill?" and "Steer Clear of Steere!" For all his efforts on behalf
of patients, Steere was now fearful of them, beaten-down, displaying a
"slightly ghostly" pallor. His stalkers thought they had a chronic form
of Lyme infection, the article noted, but Steere, "the world's foremost
expert," demurred, saying they suffered chronic fatigue, mental illness,
or fibromyalgia.

The nascent movement that questioned Lyme disease treatment, of
which the attack on Steere was part, was fueled by a novel, evolving,
and effective way to organize dissent: the Internet. Steere was a primary
focus of complaints that sometimes turned ugly. This much was true.
But for every patient who may have sent a threatening email—the article doesn't state how many Steere received—there were many more who
were doing the real work of organizing a crusade. "The group did hold
up signs and chanted things like 'Steer Clear of Steere,'" said a woman
with a cane who handed out leaflets on Fifth Avenue in New York.
"This was a group of ill people, along with their friends and family, that
were out there to educate others about the controversies surrounding
Lyme Disease," she wrote in a blog. Published letters to the editor—the
Times said it received a "flood" of mail—ran four to two against the article with one writer asking, "Would you describe people who are afflicted
with H.I.V., epilepsy, diabetes, heart disease or cancer as 'stalkers' if they
protested the loss of their medication?" Among the pro-Steere writers
was his good friend since age eighteen, the violinist Itzhak Perlman,
whose daughter had been misdiagnosed with Lyme disease; Perlman
called Steere an "outstanding physician." Nonetheless, in a portrayal
with real and lasting ramifications, the misguided acts of some—no
doubt unsettling to Dr. Steere—were cast as a guerrilla war of the many.

The venue in which the story was told—in America's paper of
record—validated the image as an accurate representation of chronic
Lyme sufferers and of their unsubstantiated claims. Physicians, already
told that antibiotics were curative despite indications otherwise, were

given license to dismiss and ignore long-term sufferers. Other news outlets could feel comfortable following suit. Seven years later, *American Medical News* ran an article on the controversy over Lyme disease that continues to this day. "I have observed among infectious diseases fellows that they don't want to see these patients," Dr. Paul Auwaerter, a Lyme traditionalist, was quoted as saying. "It has become a poisonous atmosphere." The fault is not solely among practitioners who rebuff Lyme cases. Patients insist they remain sick. Physicians do not know how to treat them. And the experts and public health officials who advise them have given neither permission nor tools to try. Chief among those tools is a clear-cut way to diagnose Lyme disease.

Eluding Capture

In May of 2016, the Canadian government sponsored a conference on Lyme disease in Ottawa that brought together a rare coterie of Lyme disease players: patients, activists, general practitioners, field researchers, data collectors, and most significantly, a smattering of prominent physicians and scientists on both sides of the parallel universes of Lyme disease. A rare sight at any conference that featured grumbling patients was Dr. Raymond Dattwyler, a New York Medical College professor of microbiology with a double pedigree in Lymeland. He was second author of the 2006 Lyme treatment guidelines of the Infectious Diseases Society of America, the ones that have dictated care in the United States and worldwide and been used to discipline doctors who practice outside them. As significant, he was a member of a US Centers for Disease Control and Prevention panel that met in Dearborn, Michigan, in 1994 and, as he put it at the start of his lecture, "wrote the two-tier guidelines."

"And," he continued, "I'm going to actually tell you some of the problems with that right now."

Two-tiered testing for Lyme disease diagnosis is so named because it involves two sequential blood tests. The first measures antibodies to see if a patient's immune system has produced enough to indicate potential exposure. Then, if the first test is positive, a second one, called the Western blot, checks to see if antibodies bind to specific *Borrelia burgdorferi* proteins, producing smudges on a test strip, called bands. These bands also indicate exposure—but with a catch.

A low-key fellow with spectacles, cropped white hair, and red striped tie, Raymond Dattwyler then explained in eighteen minutes the flaws in the blood tests that have long defined who is and is not infected with Lyme disease. His manner was matter-of-fact, as if everyone had known this for years. The technology was based on cultures that missed distinctive proteins, he said, while including others not specific to *B. burgdorferi.* "That criteria that was developed in the 1980s and the 1990s— there's a lot of problems with that." It is "not that good." You can be seropositive for the rest of your life, it doesn't mean you're infected."

I have attended other conferences and conducted many interviews in which the flaws of the Lyme diagnostic were similarly vented. But the speakers were mainly patient advocates and, especially, Lyme practitioners—doctors who have been disparaged by guidelines adherents for making Lyme diagnoses in the face of tests that had come back negative. Dattwyler's comments, however, were coming from the side that had designed Lyme disease testing.

Under the diagnostic rules set by the panel on which he served, and adopted by the CDC in 1995, Lyme patients must achieve a minimum number of bands to test positive on the Western blot test—two out of three for one type of antibody or five out of ten for another. Get four of ten and, sorry, that is a negative. Yet the blot is curiously constructed, even arbitrary, designed to detect certain proteins while leaving out others that may be important. For one, it is skewed toward detecting one manifestation of Lyme disease over another. The five-of-ten-band scenario was modeled on a 1993 study showing the bands correctly diagnosed 96 percent of arthritis-related cases—but just 72 percent of neurological cases. The blot also includes a band that detects the flagella, common to many bacteria, along with bands that are far more indicative of *Borrelia burgdorferi,* something like a trunk on an elephant. Lyme doctors in the United States sometimes rely on these significant markers to diagnose the disease, ignoring the need for two of three or five of ten bands when symptoms and clinical judgment suggest Lyme disease.

When Dattwyler spoke, it had been two decades since the choices were made of what bands to include, what to leave out, and how many were needed for diagnosis, decisions that had immense consequences for multitudes. Here, at this conference in Ottawa, was one of the most powerful directors of Lyme policy and practice in the United States and the world agreeing the technology was flawed, old, in need of replacement. "We didn't have good definitions of what was in those Western blots," he told the gathering. "Those were just bands on a gel."

So what do those bands, or rather their absence, mean in real life? For a boy, sixteen, living near Boston, they meant seven weeks in a psychiatric hospital. After relapsing from a previous bout of Lyme disease, the boy had been ruled negative for Lyme disease after registering four of the requisite five bands on the Western blot—as close to positive as humanly possible. Although Lyme disease is well known to cause serious psychiatric symptoms, doctors diagnosed the boy with "pure mental illness," not Lyme disease, according to an article in *International Medical Case Reports* journal. Enter a pathologist from New Milford, Connecticut, named Sin Hang Lee who had devised a test using DNA sequencing to search for the pathogen in human blood. When he submitted his findings on the boy's blood to the GenBank repository of genetic sequences of the National Institutes of Health, they perfectly matched the DNA footprint for *Borrelia burgdorferi.*

But as often happens with research that bucks traditional Lyme dogma, Lee's report was criticized in the scientific literature. While his test found *Borrelia* DNA, he was unable to culture the organism, leaving the case, as a 2017 letter in *JAMA Internal Medicine* put it, "unproven." Beyond this, however, even if the boy had been positive in two-tiered testing—and this is a comment on the fallibility of the technology itself—he likely would have been ruled a "false positive," with antibodies showing up from his previous infection. In short, Lyme testing is often a lose-lose proposition. Test negative and get no treatment. Test positive and get no treatment if already treated. This incenses Lee, a feisty former Yale professor who had escaped communist China in 1961 and is angry over the use of twin tests that miss many cases. What are the odds that the DNA he found in the boy's blood wasn't *B. burgdorferi* after hitting a match in the NIH GenBank? "It's mathematically almost impossible," he said.

Several months after the Ottawa conference, I reached out to Raymond Dattwyler, who is a tell-it-like it is kind of guy from the Bronx, where he was raised by working class, high-school educated parents, of which he is rightly proud. He was direct in his criticisms of the two tiers of Lyme testing, how they have been used, and said that they must go. "They were a stopgap measure. Those were never supposed to be cast in concrete. [They were] supposed to be used until something better came along," he told me. "Twenty years ago, I would've said they're fine. Now I say, 'oh shit, we were wrong.' It doesn't look as good as we thought it was."

When I questioned the upshot of this flawed instrument—what it meant for sick, undiagnosed patients—Dattwyler lapsed into the qualifiers guideline authors have used to simultaneously acknowledge the testing regimen's flaws while defending its use. "The biggest problem is not sensitivity," or too many false negatives, Dattwyler told me. "It's specificity—too many false positives." The major flaw, in other words, was not the negative tests among people who actually had Lyme disease, although that is certainly a problem if you are one of them. The real problem—the one Lyme researchers have been far more concerned with, as I wrote in chapter 4—were people who were wrongly diagnosed with Lyme disease and treated with antibiotics when they did not have it. The false positives. The double test, the high bar, the bright bands on a Western blot—these were all designed to avoid just that, weeding out people early on who might have this or that antibody but not really Lyme disease. But what about all those missed cases, I asked, the people who did not manifest the typical rash and tested negative? "You miss the early thing," Dattwyler said bluntly, "because your tests suck and not everybody gets the rash and doctors don't realize the rash is variable."

"But later," he said, "you don't miss many at all."

The Upshot
In April of 1996, the New York State Department of Health wrote a letter to the CDC about its concerns over the new two-tiered technology. Agency officials had gone back and reviewed their Lyme disease cases from before two-tiered testing was adopted to see how the new criteria for diagnosing and counting Lyme disease cases would play out. Officials were concerned. "If we followed a case confirmation scheme which incorporated the new two-test requirement for serologic [blood test] confirmation on our 1995 cases, 1,237 cases would not have been confirmed." That meant that 31 percent of all diagnosed cases in 1995 would have been ruled negative. The letter cited one case in particular: A patient had tested positive on the first tier and negative on the Western blot—a CDC negative overall—but had a form of facial paralysis that is a signal indicator of Lyme disease. "Do I confirm the case...?" the letter writer asked. Over time, the answer became crystal clear: You don't.

The warring camps and hunkered-down mentality that have dominated Lyme disease are a function of the diagnostics that Dr. Dattwyler spoke so frankly of at Ottawa and to me. Indeed, the major issue driving the Lyme controversy for two decades has been the lack of a dependable test to determine if someone is currently infected with Lyme disease. A 2013 Virginia law mandated that doctors inform potential Lyme patients, "current laboratory testing for Lyme disease can be problematic and standard laboratory tests often result in false negative and false positive results." Even when it works, the test indicates only the presence of antibodies—which can last long after a prior infection—and not of the pathogen itself. That glaring gap in the Lyme diagnosis paradigm has hurt patients who need care and, beyond this, hampered research: How can we reliably enroll patients in studies, know if antibiotics work, and chart the effects of treatment if tests fail in a portion of cases?

The better question might be why two-tiered testing has been so fiercely defended for so long, why its square pegs have been jammed into round holes. In 2012, I interviewed a leading Lyme disease researcher-physician who has long been allied with the Infectious Disease Society of America side. The researcher said, but later asked that I not use, this rather innocuous quote in regard to the test regime: "I don't think there's any question that everybody would like to have something better." In the world of Lyme disease politics, I learned, there was a distinct aversion to stepping outside the company line, which holds that the test is fine. Barbara Johnson, a CDC microbiologist with close ties to IDSA Lyme leaders, wrote this in a book chapter in 2012: "An extensive peer-reviewed scientific literature supports the rationale for and performance of two-tiered serological testing."

That statement works only if one believes the Lyme diagnostic's low accuracy—about half of tests are correctly positive at all stages—is normal and acceptable. This is a view the CDC has long embraced. "During the first few weeks of infection, such as when a patient has an erythema migrans rash," it has officially proclaimed, "the test is expected to be negative." The body simply hasn't produced enough antibodies. But the false negatives are okay, the CDC has held, because Lyme disease can be diagnosed based on early symptoms or by the Lyme disease rash. There are two problems with that.

First, a Lyme diagnosis is so controversial that many doctors want proof before treating. At three different points, the IDSA treatment guidelines advise physicians not to treat potential Lyme patients who do not have a rash or a positive test. In cases involving early neurologic, arthritic, and cardiac symptoms, the guidelines say symptoms simply "are too nonspecific to warrant a purely clinical diagnosis." Confirmation, they say, requires "laboratory support" or "serologic testing." This is an unambiguous way of telling physicians not to use their judgment, even in the face of symptoms and likely exposure.

Second, the CDC's study of 150,000 patients found the rash in 69.2 percent of cases; officially, the CDC maintains 70 to 80 percent of infections manifest it. But even a rash does not guarantee correct diagnosis since it may not look like the classic reddish "bull's eye" with a clear center. CDC photos show six variations, among many, including with a "bluish" hue, a "central crust," and "dusky centers." Just 9 percent of ninety-five people who developed Lyme rashes had the true bull's eye, according to a 2002 study in the *Annals of Internal Medicine*. At Johns Hopkins School of Medicine, researchers studying 165 early Lyme patients reported good news and bad: 87 percent actually had a rash—higher than the CDC estimates—but about a quarter of those were still initially misdiagnosed. Yet without this misnamed, sometimes misidentified, and often overlooked skin lesion, the guidelines insist on a positive test before diagnosis.

This is what happens in the real world. Because just 30 to 40 percent of tests are correctly positive in the early weeks of infection, because the rash is unpredictable, and because Lyme symptoms are common to other maladies, a share of people leave their doctors' office undiagnosed and untreated. Some go on to feel better, get on with their lives, and suffer crippling problems later on. That's the Lyme progression when it goes untreated. Recall that 10 to 20 percent of *early treated* patients suffer lingering problems. Most tragically, doctors have been encouraged, in cases with no rash, to allow infections to fester, then test later, even though patients may have symptoms and ticks may be active and infected locally. In patients without the rash, wrote Lyme pioneer Allan Steere and colleagues in 2016, "manifestations of Lyme borreliosis are typically diagnosed by recognition of characteristic clinical signs and symptoms along with serological testing." The keywords in that sentence: *along with*. Diagnose by symptoms, Steere is saying, but also have a positive test. The assertion was made in a review of the literature published in *Nature Reviews Disease Primers*, one of many recitations of previous studies that have hammered home the mainstream Lyme message.

The CDC's laissez faire pronouncements, its reassurances that a faulty technology works, and the advisories of the most esteemed names in Lyme disease, I'd argue, have made doctors complacent, believing, wrongly, that either a rash or, sooner or later, the twin tests will diagnose their Lyme cases. In fact, neither can be counted on to occur, most especially early on but later too. Further, reassurances that the tests work have stalled urgently needed research. If it isn't broke, as the saying goes.

Roberta L. DiBiasi, a pediatrician, wrote somewhat more realistically on the tests than CDC's Barbara Johnson—if in somewhat dry medical prose—in a 2014 article in *Current Infectious Disease Reports*: "Many attempts have been made to evaluate serologic testing" for Lyme disease, she stated. "For even this basic measure of test validity, there is marked controversy in the medical literature."

When I began reporting on Lyme disease in 2012, I asked Gary Wormser, the lead author on the Lyme disease guidelines and the physician most associated with Lyme policy in the United States, if the tests worked. It was the last time he would speak with me. I subsequently wrote an article that questioned the validity of the tests. He said then, in a comment that captures the one-hand, other-hand nature of two-tiered Lyme diagnosis: "We don't recommend testing for people with the rash. A negative test doesn't prove anything. If you're sick six months, six years and you don't have a positive test, give me a break." This is the prevailing principle of Lyme diagnosis: The tests don't work early, but most certainly work later. No rash, no positive test, no Lyme disease. What's the issue?

Under this regime, nonrash patients with equivocal symptoms, such as flu-like illness, headache, and fever, may be told to return for testing if symptoms persist. Yet even then, cases may be missed. A CDC continuing education tutorial advises doctors that "convalescent phase" patients, the second stage after acute, will correctly test positive in standard two-tiered testing just 26 to 61 percent of the time—the range of four studies quoted that demonstrates the tenuous nature of Lyme diagnosis. Later on, patients with "early disseminated" Lyme disease, with symptoms like meningitis and facial palsy, the four studies reported, will be positive 73 to 88 percent of the time. That's better but misses potentially one in four cases. It isn't until the "late disseminated" phase that two-tiered testing reaches accuracy heights of 95 to 100 percent, the tutorial advises. Yet those are some of the toughest cases to treat.

In 2016, British researchers looked at eighteen published studies and found the tests correctly positive just 54 percent of the time overall, a low figure that reflects early failure rates. Notably, these researchers found no standard definition of each Lyme stage—early, late, convalescent—which, they said, "prevented clear evaluation of test sensitivity." When they looked at results by manifestation, they found good results in arthritis cases—96 percent accuracy. But for neurological Lyme disease, which can lead to memory and cognitive problems, numbness in the extremities, or psychiatric disorders, the study said testing was correctly positive in 87 percent of cases overall, leaving a significant share of potentially impaired people undiagnosed.

Beyond this, studies that measure the accuracy of Lyme disease tests should be viewed skeptically. Some rely on a kind of circular logic, selecting patients on which to validate the tests who have been known to suffer Lyme disease—precisely because they had already tested positive in two-tiered testing. Researchers writing in *Clinical Infectious Diseases* in 2008 acknowledged this flaw: "It is problematic to determine the frequency" of positive tests in cases involving neurologic, cardiac, or joint problems because positive testing is "a part of the case definition." A CDC-led study also acknowledged, "the possibility of selection bias toward reactive samples cannot be discounted."

These and other flaws became eminently clear when a team led by Mariska Leeflang, a Dutch epidemiologist and testing expert, reviewed the methodology behind seventy-eight studies on the efficacy of Lyme disease tests. Leeflang's 2016 article in the journal *BMC Infectious Diseases* concluded that every one of those studies suffered from "a high risk of bias" in at least one of four categories. In the end, her team's exhaustive review did not find "sufficient evidence" to endorse current Lyme diagnostics. "These [study] designs are very likely to overestimate sensitivity and specificity," Leeflang told me – namely to inflate the test's ability to predict positive and negative results. In other words, the performance rates are best-case, not real-world, figures.

Then there is how this all plays out in fast-paced labs for a technology the CDC study called "complex (and) technically demanding." Dutch researchers looked at the performance of eight commercial versions of the first tier test and five of the second in 2011; they reported "widely divergent" results depending on which combination was used. In comparison to Lyme tests, antibody testing for HIV infection is a breeze. In 2017, two British researchers calculated the statistical probability of accuracy for each test in a head-to-head comparison. For late-stage Lyme disease— when two-tiered testing supposedly works best—Lyme tests falsely ruled patients negative 17 percent of the time. The HIV test was falsely negative in 0.1 percent of cases. Writing in the *International Journal of Medicine* in 2017, the researchers noted that was a 170-fold difference.

The Lyme controversy would cease to exist if there was a better test and, moreover, a standard, predictable, and accepted way to culture

Borrelia burgdorferi, namely to grow it from a sample of an infected person's blood. To really know how well the test works, Leeflang told me, you need to measure test performance in people who are known to have Lyme disease and, as important, in people who don't. "You need a gold standard," she said. In 2013, a Lyme disease researcher at the University of New Haven in Connecticut, Eva Sapi, introduced a culture test, her results published in the *International Journal of Medical Sciences.* Sapi's methodology was challenged in an article by CDC's Johnson, who has published on, and has patents for, Lyme disease diagnostic technology. The CDC subsequently recommended against use of what I'm told was an imperfect study but a promising technology.

Instead, the agency has for many years upheld use of a diagnostic strategy that is indisputably flawed. "Until we can separate the infected from the uninfected and the cured from the uncured," wrote Elizabeth Maloney, a physician and author of the International Lyme and Associated Diseases Society guidelines, "arguments over diagnostic and therapeutic approaches will continue." Dattwyler hopes not. He has been working on a new antibody test that, if approved and marketed, will more accurately diagnose infections. He acknowledged it would not distinguish current from past infection, as with the old test, which is a huge problem in areas where people are repeatedly infected. Beyond this, he volunteered that the test on which he has worked for a dozen years would likely make him some money. "It is going to change," he said. "It's going to change because I'm one of the guys leading the charge to change it."

Catching the Strains

Joanne Drayson, a silver-haired civil servant from Merrow, England, told me her Lyme disease story across a quiet table in East Horsley, Surrey, a village southwest of London. Like others with long-lasting Lyme disease, she was riled over her treatment missteps and determined to change things for those who would follow. This is how a movement is building around the world, one Joanne Drayson at a time. From 2003 to 2007, Drayson had seen eight doctors and been told she suffered a variety of ailments including fibromyalgia, musculoskeletal disease, and myalgic encephalomyelitis, also called chronic fatigue syndrome. Her joint pain and weakness persisted even after twenty months on steroids, which suppress antibody production and may explain why she tested negative for Lyme disease. By happenstance, her real problem became clear when she developed a chest infection and was prescribed amoxicillin. Finally, four years into her search and within twenty-four hours, she felt better, much better.

"Perhaps," her primary care physician suggested, "you have Lyme disease." To Drayson, a gardener, grandmother, and soft-spoken pensioner, this was an ironic gift. In a survey, 52 percent of late-stage Lyme disease sufferers in the UK said it took more than two years to get their diagnoses. So did 61 percent of 6,000 such patients in a survey by the US-based research and advocacy group LymeDisease.org. At the time, Drayson knew nothing about Lyme disease. Looking back, she recalled two rounds of tick bites, two rounds of flu-like symptoms, and each time, rashes on her legs. She had even reported the first rash to her doctor, and it was in her records. Finally, at this juncture in 2007, Drayson was delighted to have an answer and to have gotten better on antibiotics. The problem came later, when she needed regular rounds of amoxicillin to manage her symptoms.

In the United Kingdom, as elsewhere, doctors risk raising red flags with the authorities if they use too many antibiotics in contravention of accepted guidelines for the treatment of Lyme disease. After being on antibiotics for eight months, Drayson, then fifty-seven, was referred by her primary care doctor to a rheumatologist in London for a consultation. She told the specialist that the antibiotics were helping her and she had less pain than she had had in years. She said she believed the correct diagnosis had at last been found. She told him she had read up on Lyme disease on the Internet, at which point the rheumatologist "wagged his finger" and demurred, Drayson said. The physician followed up with a letter to Drayson's primary care doctor that sums up the experience of many Lyme patients I have interviewed. "My feeling is that there is no evidence that this patient has been bitten by an *Ixodes* tick nor has she developed Lyme Borreliosis," he wrote. Instead, some past infection—he did not specify—had caused chronic fatigue syndrome in Joanne Drayson. Moreover, he wrote, "the patient now has Lyme neurosis due to her reading around the subject." He recommended that Drayson get off antibiotics, take two prescribed antidepressants, and sign up for cognitive behavioral therapy, a form of psychotherapy that helps people change the potentially negative ways they think. Drayson, the doctor wrote, "of course…was unable to accept" this advice.

For some reason, this esteemed professor of rheumatology felt more comfortable diagnosing Drayson with chronic fatigue syndrome—what the Mayo Clinic describes as "a disorder characterized by extreme fatigue that can't be explained by any underlying medical condition." Essentially, he opted for a vague constellation of symptoms rather than an illness, Lyme disease, for which there was at least circumstantial evidence, namely her recorded tick bite, rash, and symptoms. Moreover, he felt better recommending antidepressants, which, though generally safe, come with suicide warnings. He was going by the book, which rejects long-term antibiotic treatment.

In LymeDisease.org's survey of patients whose Lyme diagnoses had been delayed, "mood disorder" was the most common diagnosis they received, reported by 59 percent. The second and third leading diagnoses were chronic fatigue syndrome, 55 percent, and fibromyalgia, 49 percent. An emergency room physician from the United Kingdom, who I met at a conference in Paris, said he began treating Lyme disease with natural methods as a sideline after seeing many patients who were looking for help. "They've been through the mill of the conventional health service," Andrew Greenland said. They're told, "it's in their head."

As in any disease, some people may think they have Lyme disease but don't. At the same time, Lyme mimics many diseases to which its symptoms are often attributed. Twelve percent of LymeDisease.org's survey respondents said they had been diagnosed with multiple sclerosis before being treated for late-state Lyme disease; a Polish study reported in 2000 that multiple sclerosis patients had twice the rate of positive tests for Lyme antibodies than other patients with neurologic diagnoses. Could some MS cases actually be Lyme disease? Similarly, patients at a psychiatric hospital in Prague had twice the rate of Lyme disease antibodies than healthy subjects in the population. "In countries where this infection is endemic," the Czech researchers wrote in the *American Journal of Psychiatry* in 2002, "a proportion of psychiatric inpatients may be suffering from neuropathogenic effects of *Borrelia burgdorferi*." For such patients, antibiotics might work better than antipsychotic and antidepressant medications. If nothing else, this and research I'll discuss in chapter 7 suggests doctors should rule out Lyme disease.

In 2016, Kris Kristofferson, the legendary musician, made an announcement. After suffering for years with debilitating memory loss that had been attributed to Alzheimer's disease, he had been diagnosed with, and successfully treated for, Lyme disease. Joint pain and heart issues resolved; his memory came back.

The failure to recognize Lyme disease, and the tendency to call it something else, goes back again to the tests. Joanne Drayson, like other late-stage patients, did not test positive for Lyme disease though conventional wisdom dictates that, in her late stage of infection, her immune system should have mounted enough antibodies and compiled enough bands to signal a past, if not current, infection. But should she have tested positive, at least using conventional tests? This is where we get into how Lyme disease is viewed—a single, straightforward disease with a reliable test—as opposed to what it often is: a manifestation of one or more tick-borne pathogens that come in many disease-causing species and strains that may or may not be tested for and for which adequate tests may or may not exist. In Scotland, for example, a group of Inverness microbiologists retested archived blood and found more patients were Lyme positive based on markers for local *Borrelia* strains rather than for the strain targeted by standard two-tiered technology. These were patients who, like Joanne Drayson, had been told they did not have Lyme disease. Then there are the other infections, like babesiosis and bartonellosis, for which doctors often do not test and are not trained to treat. Indeed, science has only scratched the surface on coinfections—does one pathogen mask another's diagnosis?—and most certainly has not cataloged all the variations of tick-borne disease.

Sue Faber of Burlington, Ontario, knows the pitfalls of testing. She suffered for years, beginning in 2001, with migrating body pain, fatigue, a chronic cough, and cognitive problems so concerning that she resigned as an emergency room nurse, fearing she didn't have what it took anymore. Seven or eight specialists later, Faber's primary care doctor was stumped. "Sue, we've tested you for everything," she said. Nonetheless, the physician said she would try again and, this time, added a Lyme test. Faber flunked: positive on the first tier, negative on the second. Faber did her own research and came to believe her symptoms and progression aligned all-too neatly with Lyme disease. What ensued was an odyssey, shared by many patients, of phone calls, faxes, fights for repeat tests, rejected requests for antibiotics, and costly trips to the United States for care. An alternative test that showed her to be Lyme positive was dismissed. "She was adamant that I didn't have Lyme," she said of her doctor.

In the end, Faber, then thirty-nine years old and on disability, prevailed in a hollow victory. In January of 2016 came these words from an infectious disease physician who ran Faber's final hard-fought test: "Your Euro Lyme is positive." She did indeed have Lyme disease, but she carried a species that did not show up on the standard North American test. Faber had traveled extensively in Europe as a young adult; she wasn't the first Canadian to test for a European species. In 2007, a twelve-year-old girl was hospitalized in Toronto with back pain and facial palsy that strongly suggested Lyme disease. After a savvy doctor did a bit of laboratory sleuthing, the girl was found to have picked up a European Lyme species during travel to France, according to a case report in the medical literature. The paper warned doctors: Look for European Lyme bugs, which specifically must be requested when ordering tests.

Soon after her diagnosis, Sue Faber's story took a perplexing twist. Her middle daughter also tested positive for the European species. But, unlike her mother, she had never been to Europe, raising questions science has yet to answer. Was she infected in Canada with a recently settled European bug? Or, as some researchers and physicians think possible, was the infection passed in utero? Of note, in 2006, a European Lyme disease strain turned up on the Newfoundland coast in a colony of nesting seabirds, typical of a phenomenon playing out around the world. Migrating birds are cargo holds for ticks, as will been seen in chapter 10.

Canada is a new frontier in tick-borne disease, as disease-ridden ticks ride a wave of climate change that is projected to take them 150 to 300 miles farther north by 2050. In just four years through 2013, cases in Canada grew twelvefold, with patient groups mushrooming side-by-side with ticks. When the government's plan of action was released in 2017, the fruit of the Ottawa conference I mentioned earlier, patients found it so weak and unresponsive that a petition to scotch it drew 20,000 signatures in a week. This kind of response, like the bull's-eye rashes on children's thighs, is symptomatic of a burgeoning problem, one that government and medicine, on both sides of the border, have yet to address.

In 2012, four Canadian scientists assessed the nation's Lyme toll in an article in *Open Neurology Journal* entitled "Evolving Perspectives on Lyme Borreliosis in Canada." In it, the researchers considered questions that seem almost naïve by publishing standards for Lyme disease science. Didn't they know, I thought as I read the article, that those issues were long ago dismissed and not open to debate in the United States? Yet they remain at the heart of how to manage and contain a worldwide epidemic. The scientists questioned whether US-designed two-tiered testing worked for Canada, which has some genetically different variants of *Borrelia burgdorferi*. That diversity, a 2011 article in *Applied and Environmental Microbiology* stated, "has consequences for the performance of serological diagnostic tests and disease severity." In other words, cases could be missed. The scientists also challenged the way that endemic areas were being strictly defined, which would also cause cases to go unrecognized, untreated, and uncounted. They pointed to signs of impending danger. Dogs, as sentinels, were turning up across the country with *Borrelia* in their bloodstream. Growing numbers of infected ticks and tick species were being found other than those officially recognized.

In 1992, the paper recounted, two Canadian physicians had noted an increase in Lyme disease referrals at a clinic in Vancouver, British Columbia. At that point, the only area of the country known to be endemic for the disease was in Long Point, Ontario, some 2,400 miles away. The doctors studied sixty-five cases and concluded that just two patients likely had Lyme disease, with both illnesses contracted in travel outside the province. The other sixty-three patients, the doctors wrote, had skin diseases, rheumatological and neurological issues, psychiatric illnesses, fibromyalgia, chronic fatigue syndrome, and so on. The physicians' article in 1993 mirrored a wave of studies, referred to in chapter

4, on alleged overdiagnosis of Lyme disease in what the Canadian doctors said was an "area of nonendemicity"—namely where there were no infected ticks or people. It would be just one more year, in 1994, before the Lyme disease pathogen was reported in British Columbia and another year, 1995, before eighteen areas of the province were declared endemic. Today, Lyme advocacy groups there, as elsewhere in Canada, are still fighting for recognition. "Canadian statistics for Lyme disease are very misleading and, to me, totally unreliable," said John D. Scott, perhaps the nation's leading tick researcher. "Public health [officials] do their damnedest to discount and dismiss the problem."

To Janet Sperling, an author on the critique of her country's response to Lyme disease, Canada isn't asking the basic questions. It is merely following the lead and dogma of American medicine. Debate is stifled. Divergent opinions—those expressed at the Ottawa conference but missing from the proposal that grew out of it—are ignored. "The new action plan for Lyme disease," she told me, "is neither new nor an action plan."

Patients the Tests Leave Behind

When Brian Fallon embarked on a clinical trial of antibiotic retreatment for Lyme disease patients who remained ill, he struggled to find suitable candidates to study, even in a country with then perhaps 200,000 cases a year. Fallon, director of the Lyme and Tick-borne Diseases Research Center at Columbia University Medical Center in New York City, had to review the case files of 3,368 Lyme disease patients before his study, published in 2008, enrolled just thirty-seven—roughly one of every hundred reviewed. Significantly, more than half of patients were rejected because they didn't register true positives in two-tiered testing even though they had already been diagnosed with the disease. To be sure, only provable cases should be included in important research. But the figures demonstrate the elusive nature of finding certifiable Lyme disease with standard tests. Jill Auerbach, a long-term patient from New

York, was rejected for the study, scoring different numbers of bands—same blood draw, same lab—that combined would have been enough to signal infection.

Indeed, the most famous Lyme disease treatment trial, published in 2001 in the *New England Journal of Medicine*, included fifty-one patients who similarly flunked the Lyme disease test. For the study, they were counted as having had Lyme disease because they had had the distinctive rash. But, despite persisting symptoms, they stubbornly refused to test positive. They were, in medical parlance, "seronegative."

Sixteen years since that seronegative study charted the future of Lyme disease care, Christian Perronne is still rankled. An infectious disease physician practicing just outside of Paris, Perronne rejects the French medical guidelines for treatment of Lyme disease, which, he says ruefully, were "exactly copied" from those in the United States. The guidelines sharply limit antibiotic courses based on the *New England Journal* study, which ruled retreatment was ineffective and risky. Perronne is particularly galled by a huge incongruity. The crafters of Lyme disease guidelines contend that Lyme disease tests work and should be relied on to make diagnoses. But when it came to conducting the seminal study on treatment, fifty-one patients were included for whom the tests failed—something Lyme doctors say regularly occurs. "They'd kill me for that," Perronne told me in a call from his office at the University of Versailles Saint-Quentin. In it, he described a twenty-year battle to help a patient population that nobody wanted in a medical milieu that believed Lyme disease was "benign" and the test for it "perfect."

Seronegative is a rejected word in the modern debate over Lyme disease, at least on the side that believes the disease is easily diagnosed and the pathogen, killed by limited antibiotics. But the notion that actively infected people might test negative wasn't always anathema. In 1988, the *New England Journal of Medicine* ran a different article, entitled "Seronegative Lyme Disease," with Raymond Dattwyler as the lead author. Dattwyler and five coauthors had studied seventeen patients who, after early treatment for Lyme disease, had developed cases the paper called "chronic," another term that has been soundly rejected. The patients were infected, the researchers believed, even though their test results were no different from normal controls. "The presence of chronic Lyme disease cannot be excluded by the absence of antibodies against *B. burgdorferi*," they wrote. Instead, they cited other immune responses in their patients—similar to what nontraditional Lyme doctors do—as "evidence of infection."

The divergent paths taken by physicians on that paper says a lot about the Lyme debate. When I asked Dattwyler about the seronegative concept, he disavowed the paper as a product of a naïve time in Lyme history. "We didn't know all the nuances that were built into that stupid organism," he said. But two of the other authors, David Volkman and Benjamin Luft, still believe it to be true: some people with the bug don't test positive. I told Luft of Dattwyler's dismissal of the paper they coauthored. He said, "I don't know what the hell he's talking about."

In late November of 2016, one of the nation's leading Lyme disease pediatricians, at least according to prevailing thought, gave a by-the-guidelines talk at a New York City symposium entitled "Infectious Diseases in Children." Here, Eugene Shapiro, a balding, white-bearded professor at Yale School of Public Health, told his audience of physicians that Lyme rashes show up in lots of different ways, which is a contention of patient advocates who say atypical Lyme rashes are often written off. "Do not think a patient must have a bull's-eye rash for it to be erythema migrans," he said. Fine so far. But then Shapiro lapsed into the not-to-worry, Lyme-is-overblown themes that have characterized Lyme dogma. Don't think every rash you see is Lyme disease, he said, citing eczema, ringworm, and cellulitis as possible candidates.

But it was when he told the group the following that my mouth dropped. According a to Healio News release, "Shapiro further noted there are no symptoms on their own that should give a pediatrician reason to test for positive results." Children with body aches, headache, and fatigue usually have something else; "the vast majority of the test results are going to be false positive results," Shapiro said

In giving this advice, Shapiro, an author of the Infectious Diseases Society's Lyme disease guidelines, followed logic that has been so ingrained into the treatment culture that its flaws seem perfectly reasonable. The IDSA logic hinges on an unreliable rash to make a diagnosis. For those without a rash, it forgives tests that aren't reliable early on but holds they sooner or later will work. This logic lets the spirochete disseminate throughout the body. It holds that diagnosis is nearly foolproof in late stages. So test later. What's the harm?

When I asked Shapiro about his comments, he surprised me again. The test was not the problem; the problem was how it was used—its so-called "predictive value." The test works well on patients with a high risk of tick-bite and specific, measurable signs of illness, like a swollen knee or facial palsy, and not so well on patients with vague complaints. "I'm talking about people with symptomology that's going on for a little while," he told me. "It's a very good test later on."

This scenario, of course, leaves out children whose bodies hurt but have no rash. It leaves out patients with joint pain, depression, cognitive issues, and lethargy. The test does not fail these folks, Shapiro believes; if they don't ever show some objective sign, they simply do not have Lyme disease.

Shapiro's philosophy, as told to a roomful of health-care practitioners who would no doubt bring it back to their practices and their patients, was one reason why a faulty test for Lyme disease had been tolerated so long. Because it is the mainstream, IDSA-certified model of Lyme disease care, as published in and embraced by major medical journals. In a functional medical model, the articles in these journals, and the ones that challenge them, should be part of an evolving and nuanced science

of Lyme disease. Instead, they represent a fixed view of testing and treatment, repeated in science literature reviews in a kind of say-it-again
Lyme loop. I found two dozen such reviews since 2005, for example,
in a PubMed search of just one guideline writer's name with the words
"Lyme" and "review."

For twenty-two years, by the time Shapiro gave his talk, practitioners
have had to rely on a pair of tests that fails to diagnose multitudes of
patients and has been found deficient in numerous published papers.
The way to solve this problem, to avoid telling doctors not to test children who have a hint of Lyme disease because they may wrongly test
positive, is to get a better test.

CHAPTER 7:

An Indestructible Pathogen?

In the center of the universe of Lyme disease—orbited by the mice, deer, and birds, the vicissitudes of weather and cataclysm of climate, the chopped-up forests and spreading suburbia, and, of course, the ticks—there is, simply, the bug: The formidable, for some unlucky humans, ferocious family of Lyme disease bugs known as *Borrelia burgdorferi sensu lato*. The pathogen circulates, it proliferates, it endures like a piece of silver in a marketplace, having few earthly or social boundaries. *Borrelia* is content to nestle in a dirt-covered tick that has not fed for months, to lay low in the knee of a mutt or a thoroughbred, or to swim in the heart or brain of a billionaire. It will share space in the gut of a tick with other disease-inducing bugs, like *Babesia*, and sometimes become stronger in the process, more apt to infect than it otherwise would be. It will move from a tick to a mouse to another tick nearby, like passengers transferred between parked buses, and never infect the mouse. Nor will it cripple or kill the many small mammals that it does infect and on which it depends for life. It knows better. *Borrelia* has allowed these animals—and itself in the process—to live well and prosper. The ones to suffer are people and their dogs, horses, and sometimes cattle, all relatively new additions on the borrelial menu, and all somewhat defenseless against it.

George Poinar's tick encased in 15-million-year-old amber, when super-magnified, showed corkscrewed cells that strongly suggested *Borrelia* (chapter 3). Those coiled black shapes hint at how much time the bug has had to adapt to the tick it inhabits and the wildlife it infects, and vice versa. Indeed, *Borrelia* and tick, and *Borrelia* and host—mammals, lizards, birds—live mostly in symbiotic harmony. Were he alive today, Charles Darwin would be delighted, ecstatic in fact, to study the family tree of *Borrelia burgdorferi s.l.*, with its trunks, branches, and twigs that have blossomed over the eons and benefited so well from the process of evolution.

Like settlers moving to a new frontier, *B. burgdorferi's* modern self comes well prepared, able to colonize with a few swift moves. First, it replicates when it senses the tick in which it lives is beginning to feed. Then, it sheds a molecule on its outside called outer surface protein A, OspA for short, and dons another protein, OspC. The first molecule bound it tightly within the mid-gut of the tick, the finger-like projections that can be seen in silhouette beneath the rim of a backlit tick. Without OspA, *Borrelia* moves freely to the salivary glands—far bigger in ticks than proportionally in people—where perhaps the ultimate feat of evolutionary cooperation occurs. Two proteins, one from *Ixodes* tick saliva called Salp 15, the other from *Borrelia*, the OspC, combine—yes, join forces—to assure product delivery from tick to host, allowing the spirochete to plant its flag on mammalian turf and protecting it from death-by-antibody. A Hungarian scientist, Gábor Földvári, called this, in an understatement perhaps, "an intriguing example of co-evolution." The whole relationship is, in fact, a case study in mutually beneficial coexistence. Beyond the advantages that *Borrelia* reaps from the tick, we also know the tick wins too: Ticks that are infected are literally fatter (as in they have more body fat), more resistant to drying up, and able to search farther and longer for food. And by the way, they outlive their uninfected compatriots.

The real story of *Borrelia burgdorferi,* the one we care most about, is what happens after that clandestine tick-to-host handoff, as the organism sets up shop, multiplies, and travels throughout the human apparatus. That's what makes it so dangerous. It can move with its flagella when it wants to and stay put when it needs to. It enters the skin and circulatory system first, of course, moving to joints perhaps, or, depending often on the preference of its assorted species and strains, taking up residence in the central nervous system, the peripheral nervous system, the heart, brain, or even the eyes. This bug knows how to drill through that tight-knit helmet of cells known as the blood-brain barrier, and wreak havoc on memory, understanding, and mental health generally. Lyme disease has been associated with depression, schizophrenia, and obsessive disorders. It can bring on what one study called "pure Lyme dementia," similar to the madness wrought by another spirochete, *Treponema pallidum*—syphilis. When *B. burgdorferi* camps in the joints, it makes movement, for some, excruciating; in the peripheral nervous system, it causes numbness, tingling, or weakness. In cardiac tissue, it slowed the heart rate of a nineteen-year-old Colorado boy with Asperger syndrome to twenty-five beats per minute, causing him to fall and strike his head. He was treated and survived. A seventeen-year-old New York boy, his heart invaded too, did not.

How *B. burgdorferi* thrives in the human body is the stuff of scientific mystery, one that Nicole Baumgarth, an immunologist with a degree in veterinary medicine from Hannover, Germany, is determined to unravel. She has spent hundreds of hours in a laboratory at the University of California, Davis, in pursuit of a fundamental question. How, she asks, can this organism so fluidly and artfully evade mechanisms of immune response in ways that few pathogens do? Previous researchers had shown that a magical protein in tick saliva, when injected through the bite of a tick, suppressed invader-fighting T cells essential to human immunity and health, while triggering myriad other unsavory responses. It was also known that people treated for *Borrelia* infections responded to antibiotics in a way unlike other diseases: with a huge, rapid decline in antibodies that under other circumstances would remain and protect against future infections.

In her pursuit of the key to the borrelial kingdom, Baumgarth studies mice, many, many mice. For one, they are veritable *Borrelia burgdorferi* factories, chiefly, though not solely, responsible in many corners of the world for keeping the organism alive and circulating in the environment. For another, they are persistently, chronically infected without being sick—at least not after the initial infection. Find out how the mouse achieves this, and you may have found a cure for intractable Lyme disease.

I met Nicole Baumgarth—face to face at least, we'd already spoken on the telephone—over an urn of coffee at a conference in New York City in November of 2016. She is tall, like me, with curly red hair, gold-rimmed glasses, and an unassuming manner that says no question is too small. I appreciated this since my endless queries to many scientists, for a book that covers a great many branches of tick, mammal, climate, and germ study, can come off as, well, rather basic. Her research on mice had preceded her: a Lyme physician, among others, had pointed to it as explaining a key mechanism by which *Borrelia* lays claim to its host with magnificent impunity.

Baumgarth was introduced at the conference as someone who studies how *Borrelia* "co-opts" normal defense systems, as a "a mouse translational researcher," who is using the mouse to figure out the human. "*Borrelia*, over evolutionary times, established sort of a truce with the mouse immune system," she began. The mouse, more or less, agrees to host the bug. The bug agrees not to make the mouse sick. "Perfect balance," Baumgarth called this. In humans, however, "there is no evolutionary adaptation," she said, no *Borrelia* truce. There is, instead, all-out war.

The immune system of a mouse is not all that different from that of a human being. Under normal circumstances, tiny powerhouses called "germinal centers" are activated in the lymph nodes—human or mouse—as a key immune defense from an invading pathogen. They help to build the infrastructure that fights infection and prevents reinfection: the plasma cells that support antibody production; the memory B cells that provide long-term immunity and respond to a recurring threat. In 2015, Baumgarth and her UC Davis team reported a striking discovery. The germinal centers in Lyme-infected mice were abnormal, and the plasma and memory cells failed to develop for months after infection. In fact, the Lyme disease bug was so effective at switching off the immune system that infected mice given the influenza vaccine failed to mount the normal viral antibodies for protection, the team reported in *PLoS Pathogens.* "What has *Borrelia* done to establish the suppression of the immune response?" she asked when I spoke to her. Might it be possible to stimulate the parts of the human immune system that *Borrelia* subverts?

Dr. Stephen Barthold, a pioneering UC Davis researcher whose path Baumgarth is following, has great respect for *Borrelia,* having studied it for twenty-five years, primarily in mice. His work earned him election in 2012 to the National Academy of Medicine, among the highest honors a medical or veterinary scientist can achieve. "Once *Borrelia* gets in the door and establishes infection," he said in an interview with the Canadian Broadcasting Corporation in 2013, "it can escape host immunity 100 percent of the time."

Persistent. Ironclad. Indestructible?

People like Baumgarth and Barthold often use a particular word to describe the state in which the Lyme bacterium can live in animals and, likely by extension, in people. *Borrelia,* they contend, "persists." It leaves its host unable, in quaint scientific parlance, to "mop up" the cells that antibiotics leave behind. The ability of the Lyme disease bug to persist potentially explains why many people stay sick even after they are treated. The figures vary on just how many there are, as I discussed earlier. But consider this from a leading physician in the National Institutes of Health, Adriana Marques, from a 2012 presentation: "Studies of patients with erythema migrans [the Lyme rash] have shown that 0–40% of the patients have persistent or intermittent non-specific symptoms of mild to moderate intensity 6–24 months after therapy." Now consider that these are the people who were treated early, because they had the confirmatory rash, and who therefore achieve the highest cure rates.

These persisting cells—dormant, altered, attenuated though they may be—are still very likely infectious, Barthold believes, even after being dosed with frontline antibiotics that have long been recommended for Lyme disease. *Borrelia burgdorferi*, Barthold told a Congressional hearing in 2012, is "a professional at persisting." He reeled off five antibiotics that the pathogen had survived in treated animals, and referred to laboratories in four states and Finland that, to that point, had come to identical conclusions.

Persistence isn't an unusual phenomenon. All bacteria form persisters even after the animals they've infected are treated. Many are harmless, and the immune system can manage them. But some are not. Dormant persister cells, for example, are known to be responsible for the need for extensive antibiotic courses for tuberculosis. In 2014, a Johns Hopkins University team led by Professor Ying Zhang reported that twenty-seven antibiotics were more effective at killing persisting *B. burgdorferi* cells—some in atypical "round body" forms, others in tough-to-penetrate microcolonies—than the leading Lyme drugs doxycycline and amoxicillin. These included heavy duty antibiotics like daptomycin, used for resistant staph infections, and leprosy drugs like clofazimine; when used with Lyme drugs like doxycycline, they were much more effective at killing persisting *B. burgdorferi* cells.

At about the same time, a team of researchers at the Antimicrobial

Discovery Center at Northeastern University in Boston, led by a distinguished biology professor named Kim Lewis, similarly sought to test the pharmacological tolerance of Lyme disease bugs that had survived initial antibiotic treatment. When the researchers tried to kill *B. burgdorferi* persisters, it was a bit like playing whack-a-mole. They tried three antibiotics—amoxicillin, ceftriaxone, and doxycycline—individually and in every two-drug combination; each time, persisters survived. They tried daptomycin, known to kill pathogens rather than simply slow their growth, and, separately, several experimental compounds, and persisters survived. They turned a corner when they resorted to Mitomycin C, a highly toxic anticancer agent. At last, the persisters were vanquished. As for antibiotics, however, they finally tried a method in which ceftriaxone was applied in "pulsed" doses. Three times it was washed away, and three times the bacteria revived. On the fourth round of dosing, the *B. burgdorferi* persisters, at last, died.

The Northeastern researchers, who published their findings in 2015, and those at Johns Hopkins were among a long line of scientists that has wrestled with the Lyme disease pathogen and found it hard to kill. Their results demonstrate the complexity of the infection, which may require different treatment approaches depending on when it is attacked and the persistent form it is in. Clearly, more research is needed. When Stephen Barthold appeared before Congress, part of a hard-fought effort by Lyme disease sufferers to get official attention, he recounted twenty-five years of study into the ability of *Borrelia* to survive the best weapons of pharmacology and immunity. He had been part of Monica Embers' groundbreaking research that had tried to kill the infection in rhesus monkeys, only to attach clean ticks afterward and learn they had picked up the bug—from monkeys treated with curative antibiotics. "We can also look in the tissues of the animals that have been treated with antibiotics," he said, "and we see morphologically intact spirochetal forms." Which was exactly what he and Embers had found.

Emir Hodzic, a colleague of Baumgarth's and Barthold's at UC Davis, speaks about *B. burgdorferi* in almost reverential terms. He and Barthold had tracked infected mice for a year after aggressive antibiotic treatment, expecting the pathogen to persist in small amounts, as other researchers had found in shorter-term tests. Instead, he was stunned to learn the mice carried as many spirochetes as controls that had never received the supposed cure. "*Borrelia* is a really, really interesting microorganism," he told me. "It has a huge brain. It outsmarts people." These bacteria don't just resist antibiotics, he said. "They tolerate antibiotics."

The disheartening thing about Hodzic and Barthold's mice, however, was this: When they removed spirochetes and tried to culture them— to grow more of them in a petri dish—nothing happened. The bugs stubbornly refused to be cultured. This is where *Borrelia burgdorferi* confounds and thwarts every effort to be understood. By this point in the research, the Davis team had confirmed something significant was happening. In mice, which should have been *Borrelia*-free, they found the bacterium's DNA (indicating it had been there), its RNA (indicating it was replicating), and, nestled within mouse tissue, the spirochete itself—in fact, many of them. They had transplanted these spirochetes in tissue into other mice and infected them. They had also attached ticks to the "cured" mice, and the ticks became infected. Yet the bugs themselves—"viable, but uncultivable," Hodzic called them—would not grow outside of the mouse.

To Hodzic, this was part of the character and mystery of persistent, some might say chronic, Lyme disease infection. "Having worked with *B. burgdorferi* for over 21 years, it is apparent that not all isolates or strains can be easily cultured," he told me, "and this is especially apparent during long-term infection." Barthold referred to this dynamic in his Congressional testimony. The inability to culture the spirochete, he acknowledged, was an absence of one kind of proof. But he cited all those other ways in which presence of surviving *Borrelia* had been confirmed repeatedly. "Naysayers refute these collective observations for a variety of reasons," he told me, including study design, and how mice were infected and treated. The other thing the naysayers do, Barthold said, is "simply ignore them altogether."

Poles Apart

When I first began this line of reporting in 2012, I spent an hour on the phone with a respected scientist in the study of Lyme disease. As I had with many other researchers, we hashed out the issues that have dogged this epidemic, the chief one being whether Lyme disease can be chronic, can persist. Unlike some scientists I had interviewed, this scientist did not dismiss the contentions of those with opposing views. In the end, this scientist said, "You have to decide what side you're on." It was a frank admission and an assessment of the science of Lyme disease, a worldwide epidemic on which there were and still are clear sides. Each has looked at the science, chosen what to acknowledge and what to dismiss, and gone in opposite directions.

"I find myself in a rather contentious field," Stephen Barthold commented to that Congressional panel, "coming out of the mainstream of Lyme disease research into one in which I am somewhat of a pariah, in terms of the established medical opinion." Like other scientists, Barthold had come to believe that modern medicine had it wrong on Lyme disease. Antibiotics, he had concluded, did not always kill the bug, at least as commonly prescribed.

Linda Bockenstedt, a Yale University researcher, is the yin to Barthold's yang, an esteemed scientist with a pedigree every bit as weighty as Barthold's (who had come to Davis from Yale) but with a very different view of Lyme disease persistence. In 2002, Bockenstedt and Barthold shared credit on an early study that dosed infected mice with antibiotics and then found spirochetes in ticks that fed on four of ten of them. But disease carrying though they may have been, the ticks did not go on to infect uninfected mice. Significantly, the spirochetes were damaged, lacking some of the basic genes related to infectivity. The two scientists concluded then that spirochetes lingered to be sure, but, of paramount importance, the changes in them meant they were not infectious. By 2012, Bockenstedt, working again with mice, was convinced of this; Barthold, working on experiments on monkeys and mice, published in 2012 and 2014, of just the opposite.

To be sure, something lingered in the joints and surrounding tissue of diseased mice that had been given antibiotics, Bockenstedt reported in a study published in the *Journal of Clinical Investigation.* But they were not viable, infectious organisms of the kind that lived in the mice before treatment. Rather, Bockenstedt's team found a kind of debris, seen in fluorescent deposits near ear cartilage and within joint tissue of mice, which contained bits of spirochete protein. That detritus was enough, she and her colleagues theorized, to cause "prolonged inflammatory responses in the joint after infectious spirochetes have been eradicated." In other words, people who suffered ongoing symptoms after Lyme disease treatment, in this case those with arthritis-like symptoms, may not have chronic disease but a prolonged autoimmune response to vestiges of illness.

"This study provides the first direct evidence that spirochete proteins can remain long after the bacterium is gone, and in places where people can experience symptoms after treatment for Lyme disease," Bockenstedt told YaleNews. "These symptoms after treatment may not be caused by the Lyme bacterium itself, but by an immune system that is slowly removing the nonviable remains of the long-dead bacterium."

In key ways, Bockenstedt's latest findings were not the polar opposite of her earlier ones with Barthold. *Borrelia* DNA was found in five of her twelve mice after treatment with doxycycline, enough in the view of some Lyme doctors to equal infection. *Borrelia* was even cultured from a tick that had fed on a treated mouse, though not from eleven others. Shouldn't that one mouse, representing 8 percent of the sample, have been cured?

A commentary that appeared with Bockenstedt's article said the Yale team offered "compelling evidence" that residues of the Lyme bug—not the bug itself—were "continuing to cause mischief long after bacterial life has ceased." Once again, the message was broadcast that antibiotics had done their job against Lyme infection. But, akin to research on the other side of the Lyme divide, the findings were by no means conclusive and in some ways contradictory. Yes, Bockenstedt found evidence of potentially damaging cellular debris. But she also found spirochetal DNA and, at least in one mouse, spirochetes. No matter.

"Bockenstedt et al. recognize the far-reaching consequences of claims about persistent viability," said the commentary, written by Alan Barbour, an early Lyme disease researcher. "They come down on the side of the existence of non-viable and non-transmissible posttreatment spirochetes in their mouse model." Hence, they favored the theory that remaining spirochetes could not transmit disease. In the end, the mainstream side warmly embraced Bockenstedt's findings. The disagreement, like the spirochete, persisted, though just one side prevailed.

"Fuel to the Fire"

In 2012, a study in the *New England Journal of Medicine* repackaged an oft-repeated bromide of Lyme disease: It doesn't come back after antibiotic treatment. In the article, several authors of the prevailing treatment guidelines reported that seventeen Lyme disease patients who had repeat Lyme rashes over a twenty-year period got them from another tick bite, rather than a relapse from their original infection. Repeat bites and recurrent rashes occur frequently in tick-ridden areas—I have had two rashes, two quickly treated infections—so this was not a startling conclusion. Nor was it news that new rashes, the basis of the study, supported evidence that people had been bitten and infected again.

So how meaningful was a study of seventeen patients who got two infections in twenty years? Did it disprove relapse? In 2013, I raised this question with two of the primary authors of the study, Gary Wormser and Robert Nadelman, for a newspaper article I was writing. Both were also authors of the Infectious Diseases Society of America Lyme disease guidelines.

I was told, through a spokesperson for New York Medical College, where the physicians were based, that the reinfection study had appeared in a "highly respected medical·journal" and they would have no comment for the "lay press." Ironically, the study's findings had already been widely reported in the lay press, from NPR to the *New York Times*. The college had even put out a press release on it—part of a public relations effort that has solidified the easy-to-treat myth of Lyme disease. I was rebuffed because I had questioned mainstream Lyme dogma. Similarly, I reached out for this book to Wormser and Nadelman and, after several tries, was told in an email from Jennifer Riekert in the medical college press office, "The Dr.'s have declined your invitation for an interview."

One author of the reinfection study was Dustin Brisson, a University of Pennsylvania researcher and expert in disease ecology, who did kindly respond to my inquiries. He said the conclusion used basic statistics to compute the probability of reinfection or relapse given the strains found in one infection to the next. I asked him if the study had convinced him that Lyme disease was not a chronic illness. "I am not convinced of this at all," he wrote in an email, "and it is not what the paper suggested either." Nonetheless, the paper was widely reported as bolstering the case that Lyme disease, once treated, does not recur. Eugene Shapiro, an author of the Lyme guidelines, said in a MedScape video that the study provided "very strong evidence that the recurrences are new infections as opposed to persistence of the original infection." Here was another study that disproved the notion of chronic Lyme disease—even if it only considered people who got the rash twice.

In his commentary on Linda Bockenstedt's landmark article on spirochetal "debris," Alan Barbour wrote that the Lyme debate hinged on a choice "between lifeless leftovers and viable hangers-on," namely between spirochetes dead after treatment or alive and well. After three decades, the Lyme disease "controversy," a word he used five times, lingered like pesky posttreatment symptoms. Yet here was a classic example of why so little had been accomplished to bring opposing sides together. On the side of lifeless leftovers was Bockenstedt, whose research—diligent and honest as it was—comported with long-held establishment views and was warmly received. On the side of viable hangers-on were findings by Monica Embers who had recovered spirochetes and *Borrelia* DNA from treated monkeys but whose study, a critique by guidelines author Gary Wormser said, was plagued with "important flaws in experimental design." That lone appraisal would be footnoted by Bockenstedt two years later when she wrote that Embers' findings had "evoked considerable skepticism."

This is all part of the mainstream echo chamber of Lyme disease. Bockenstedt, Barbour wrote in his commentary, provided "compelling evidence"; Embers "added more fuel to the fire." But Barbour couldn't seem to commit because, well, you never know. He ended his commentary with a study from a half-century before that had found remnants of a bug that caused kidney infection even after antibiotic treatment. "These remains in the urine were uncultivable by routine procedures," he wrote, "but, according to the authors, lived to cause disease again."

Both Embers' and Bockenstedt's papers were published in 2012. Bockenstedt's article appeared in a journal with four times the all-important "impact factor" of Embers', a measure of how often a journal's published research is cited by other scientists. This was and is symptomatic of the publishing struggles of scientists who challenge Lyme dogma. They find outlets where they can. Nonetheless, Embers' landmark research would not be squelched. It was read more than 45,000 times on *PLOS-One* and dubbed a "Faculty of 1000" paper, a distinction by a clinical research service reserved for important findings. In answer to the uproar that followed her article, Embers posted a comment on *PLOS-One*. Her goal was to undertake "objective, well-performed experiments on antibiotic efficacy" in monkeys, she wrote, rather than to challenge the hot-button issue of treatment regimens for people. Doing that would take "solid proof of better treatment options," she said pointedly. "These are currently not available." This, indeed, is another outcome of the accepted model of Lyme disease care. Research is scant on better ways to treat Lyme disease because there is little impetus to look.

Lessons of Infection: How They Hide

Garth Ehrlich knew nothing about the so-called Lyme wars, as they have been described, when he began looking into the parallels between *Borrelia burgdorferi* and other difficult-to-eradicate bacteria he had studied since the early 1990s. Ehrlich, then an associate professor at the University of Pittsburgh, had had an "ah-ha" moment in the mid-1990s when he read an article in *Science* magazine by a microbial ecologist named Bill Costerton. In the 1970s, Costeron had noticed a slimy substance clinging to rocks in pools in the Canadian Rockies with high bacteria levels, finding they were essentially protective coatings for an army of bacteria. Now, Costerton was discussing how difficult it was to culture and kill bacteria beneath what he came to call "biofilms." Ehrlich wondered: Could this explain the intractable nature of some human infections he had studied?

In 2002, Ehrlich infected forty-two young adult chinchillas, basically one-pound rodents covered in velvety fur, with *Hemophilus influenzae*, which is a leading cause of ear infections in children. When the animals were euthanized, Ehrlich made a groundbreaking discovery that was rooted in Bill Costerton's work in the Rockies. Inside the rodents' ears, beneath amorphous, impenetrable biofilms, were *H. influenzae* bacteria, uncultivable and resistant to antibiotics. Four years later, Ehrlich, by then a professor at Drexel University College of Medicine, directed a team of scientists that analyzed bacterial samples from the chronically infected inner ears of twenty-six children. Sure enough, in twenty-four of the children, pathogenic bacteria were found under layers of nonliving matter where, as in the chinchillas, they had found refuge until the time was right to emerge and cause sickness. "They build a little house for themselves," Ehrlich told MIT Technology Review when his team's article was published in the *Journal of the American Medical Association.*

In 2012, Ehrlich was asked to give a talk in Hershey, Pennsylvania, at a conference organized by a former Fortune 500 company consultant and champion of Lyme disease patients named Julia Wagner. The mother of three had also found herself, within two years of moving to the Philadelphia area in 2004, hit by virtually every punch the Lyme bug can land, including crippling joint pain, early dementia, and cardiac problems. After she at last found a Lyme specialist who would diagnose and treat her—"I didn't get any help from the medical community," she said—Wagner began to organize and agitate. She turned to Ehrlich. Ehrlich knew a thing or two about chronic disease, she reasoned, and Wagner knew that a savvy scientist named Eva Sapi had found Lyme pathogens alive and well beneath biofilms. Could Ehrlich speak at the Lyme conference? Ehrlich's introduction to Lyme disease came in fits and starts over the next eighteen months, but he soon made connections, saw parallels, and, ultimately had another ah-ha realization.

The clincher came in October of 2013, when Ehrlich listened as a researcher from Switzerland, Judith Miklossy, offered a potential bacterial explanation for Alzheimer's disease: spirochetes, and, in particular, the Lyme pathogen, *Borrelia burgdorferi.* A presentation at a conference of the American Academy of Oral Systemic Health might not be a game-changer for many people. It was for Ehrlich, who had earlier published on the link between periodontal spirochetes, namely on teeth and in gums, as a cause of chronic infections in artificial knees. "I was putting 2 and 2 together," he told me. Spirochetes could lead to chronic infections. The Lyme spirochete often spurred lingering symptoms. It was able, in petri dishes and human tissue, to form antibiotic-resistant biofilms, just as in Ehrlich's chinchillas. Periodontal spirochetes hid beneath biofilms, causing gum disease. A pathologist named Alan MacDonald had discovered Lyme spirochetes as early as the 1980s in the brain tissue of Alzheimer's patients, while Miklossy had published a score of articles in the decades since on the mechanisms behind spirochetal infection in the brain.

When scientists make such connections, they fall back on a huge chapter in medical history involving another spirochete and its link to dementia. *Treponema pallidum*, the cause of syphilis, was legendary for leading to psychiatric illness that, before the advent of penicillin in the 1940s, accounted for 20 percent of admissions to US mental hospitals. "These are some of the most famous persistent bacteria in all of human health and disease," he told me. And they are cousins of *Borrelia burgdorferi*.

Ehrlich, then in his midfifties, had another reason to be drawn to Lyme disease. His sister, Susan, who resided in Connecticut, a state that is ground zero in modern times for the disease, had had perhaps five Lyme disease infections in the previous fifteen years. They had left their mark. Growing up in the Genesee River Valley in New York State, he recalled, self-deprecatingly, "I was like the village idiot." Susan, "monstrously gifted," he said, had been valedictorian of her high school class, with a cracker-jack ability to master and recall information. When I spoke to Ehrlich in June of 2016, he said his sister "can't remember anything—she says she lives in a world of brain fog," which Susan confirmed when I spoke with her by telephone from her home in Woodstock, Connecticut.

After the Miklossy talk, Ehrlich explored the Lyme disease research, discovering, as others had before him, a contentious scientific landscape, dominated by a powerful camp that denied the bacterium survived antibiotic treatment and left little room for debate. He saw this firsthand, when he served on a National Institutes of Health panel that reviewed a grant proposal to study basic mechanisms that might allow the Lyme disease pathogen to persist. He and two other primary reviewers gave it high marks, virtually assuring it would be approved. But when presented to the entire panel, he said, an infectious disease physician spoke up. "This shouldn't be funded," Ehrlich recalled the physician saying. "Everyone knows there's no such thing as chronic Lyme disease." A second infectious disease doctor agreed, and the proposal was killed. Ehrlich was stunned.

"Why the Lyme disease field has become the most polarized field in microbiological science I don't really understand," he told me. "It is an enigma to me why people are so fervently religious in their statement that it can't possibly cause chronic disease to the point they attack their colleagues. Neither side has enough data to assure they are right." But one thing was clear to Ehrlich: "Orthodoxy has just killed research in Lyme disease."

So toxic was the notion of persisting Lyme disease that an infectious disease physician at Drexel expressed concern for the university's reputation if Ehrlich cohosted a meeting of physicians in the International Lyme and Associated Diseases Society; the ILADS group does not hew to prevailing treatment guidelines and holds that Lyme can sometimes be chronic. ILADS, with a few hundred physicians and a core of supporting scientists, is the tick in the scalp of the better-funded, better-connected Infectious Diseases Society of America. The meeting was nonetheless held, in 2015.

Strains, Species. Known and Not.

Once inside of people, *B. burgdorferi* embarks on an adventure worthy of the 1966 movie "Fantastic Voyage," in which miniaturized explorers ride a tiny submarine along the turbulent human bloodstream, through a pounding, pulsing heart, and into the body's deepest, darkest warrens. The goal: to save an important scientist's life. *B. burgdorferi* has no such life-saving mission, but its ride is no less stormy and fraught. Through the magic of cell imaging technology, scientists from the University of Toronto watched in real time as this bug latched onto proteins in cells that line blood vessels and swung, bungee-cord style, in flow chambers that simulated the human body, from cell to cell, all neat and controlled. This bacterium, unlike that tiny, tossed vessel, has learned through the process of evolution where it needs to go and how to get there. Imagine that. Each *Borrelia* species has instinct written into its DNA, depending on its genetic profile.

Within the giant family known as *Borrelia burgdorferi sensu lato*, there are the three major, but not only, species of spirochetes that cause human illness. They are *Borrelia burgdorferi sensu stricto* (in the narrow sense) in the United States and Canada, and *Borrelia afzelii* and *Borrelia garinii* in western and central Europe, Russia, and China. The Eurasian species *B. afzelii* is more apt to stay in the skin, in late stages causing painful bluish-red discoloration and swelling—called "acrodermatitis chronica atrophicans"—which may appear months or even years after a tick bite. *B. garinii*, also widely circulating in Europe and Asia, more typically leads to neuroborreliosis—neurologic Lyme disease—with its meningitis, facial paralysis, gait abnormality called foot drop, and sometimes devastating cognition and motor challenges. Going where nature calls it, *Borrelia burgdorferi* in North America, like its cousins worldwide, also uses the bloodstream not as a destination but a motorway, exiting quickly and confounding efforts to find it in blood tests. Research suggests this New World species colonizes many more anatomical corners than its European relatives, creating its own distinct havoc. Get a Lyme rash in Slovenia and it will last about fourteen days, a study in the *Annals of Internal Medicine* reported; get one in New York and it will last just four days but will be accompanied by more systemic symptoms. As a 2016 paper in the journal *Emerging Infectious Diseases* put it, *B. burgdorferi* is more often "associated with a greater number of disease-associated symptoms," including but not limited to arthritis, the disease that led to its discovery in 1970s coastal Connecticut.

Science has a rudimentary knowledge of *Borrelia's* many manifestations, with species and strains still emerging, along with evidence that they behave differently and may not be caught by diagnostic tests. Recall the Canadian girl who came back from Europe with Lyme disease and flunked the North American test, along with the Scottish blood samples that turned positive when a different test strain was used. Beyond this, species are still being identified and cataloged, for which, logically, tests don't exist. The pathogen behind a Lyme-like illness in the American Southeast still hasn't been identified. Ditto Australia, where, as one study put it, "Four published studies have searched for *Borrelia* in Australian ticks, with contradicting results." Officially, Lyme disease does not exist there, despite more than five hundred reports in the scientific literature that suggest some kind of illness like it most surely does.

The unfolding nature of infections caused by *Borrelia*, with twenty-one genospecies and hundreds of strains so far identified, have added greatly to worldwide uncertainty and the woes of patients. In China, sixty-five *B. garinii* strains and twenty-two of *B. afzelii* have been identified. In the United States, one study looked at four hundred variations of *Borrelia burgdorferi*—"the largest number of borrelial strains from North America ever to be investigated"—and identified those best at entering the bloodstream. Another US study of just fourteen strains found huge variation by strain in the ability of standard testing to diagnosis Lyme disease. Among the study's 158 infected patients, just 37 percent tested positive for the disease. When a different antibody test was used, the share of positives jumped to 68 percent. Indicative of the gaps in Lyme tests, the strains in eighteen patients weren't detected at all. Such findings make clear: strains present yet another challenge to the ability of doctors to diagnose and treat.

Nataliia Rudenko is a Ukrainian scientist who for a couple of decades had been studying *Borrelia* species at the Institute of Parasitology of the Czech Academy of Sciences, authoring dozens of papers with US and European colleagues. In 2016, she came to a Canadian government Lyme disease conference in Ottawa armed with charts and maps that together formed a census of *Borrelia burgdorferi's* growing iterations. Tall, brown-haired, and speaking rapidly in fluent English with a Slavic flair, Rudenko was eager to share her startling findings. In 1981, Willy Burgdorfer had identified the spirochete that caused the disease, and everybody thought: problem solved. But no, Rudenko said, "That was the beginning of the problem."

By 1998, there were ten known species; by 2016, twenty-one. She made it clear that science was not done counting. With that, Rudenko unveiled three progressively marked-up world maps. The first, from ten years earlier, was "very clean and very easy," she said as the slide lit up. The second, from five years before, was "kind of a mess," she said, "but, believe me,"—she clicked on her third slide—"it's nothing like it is now." Here was the globe of the moment, the known universe of *Borrelia burgdorferi sensu lato*. From one slide to the next, ovals and arrows and shaded territories blossomed. Names like *B. japonica* and *B. turdi* appeared in the Far East; *B. bissettii* and *B. garinii* showed up on both sides of the Atlantic. New species cropped up in California, the American South, and Western Europe. Finally, something called *Borrelia mayonii* appeared in 2016, in the American heartland, a nasty species able to achieve high levels of what's called "spirochetemia"— namely blood-borne spirochetes. It had been known to have sickened six people and seemed limited though, the implication of all this was, not contained.

By the time of Rudenko's talk, some four decades into what had become a Lyme disease pandemic, about ten of the twenty-one known *B. burgdorferi* species had been implicated in causing illness, a drip-drip-drip of reports in science journals that in some cases made waves, in others fell silently into the vast ocean of *Borrelia* literature. Sometimes, such reports were ignored or dismissed, seen perhaps, as outliers or poor science. Skepticism is a healthy part of science publishing, in which theories must be proved and cases replicated. But, as I have discussed before, there has been reluctance to embrace some findings by a side of Lyme science that sees the threat as exaggerated. Researchers with proven track records and significant publications have been stymied, their findings discounted, when they veer too far from accepted, mainstream Lyme theory. Her publishing record notwithstanding, Rudenko has felt the influence—sting might be a better word—of the tamp-it-down side of the Lyme divide.

In the summer of 2013, a team led by Rudenko sought to challenge long-held beliefs about Lyme disease in the American Southeast. If a tick bite there had led to a suspicious rash, it was usually deemed a case of STARI—Southern tick-associated rash illness—or what might be called Lyme-disease lite. STARI was delivered by the bite of a different tick in the *Ixodidae* family, *Amblyomma americanum*, also known as the lone star tick. The disease was caused by a pathogen that had yet to be isolated, and it did not lead to the many complications of its more ferocious relative to the north. Home-grown Lyme disease, on the other hand, was rare. At least that was the official stance.

This view was disputed by patient advocates and a key scientist on Rudenko's team, Kerry Clark from the University of North Florida, who was a leading tick-borne disease pioneer in the South. "Lyme disease is vastly more common down here than anybody has acknowledged,"

Clark told me around the time the Rudenko study was getting under way. "The frequency of the disease is way beyond the expected level, and that's the definition of epidemic." A dogged researcher who had tested perhaps a thousand human blood samples, Clark had long argued that a large portion of recognized Lyme disease cases in southern states were contracted locally and not, as officials almost always ruled, picked up in travel to the north. He had published findings of DNA from *Borrelia burgdorferi* in the blood of fifty-three Florida and eight Georgia residents, including eighteen cases in Florida of two rare species not previously known to infect human beings. Significantly, he had found the Lyme pathogen in that most prevalent of southern ticks—the lone star—including one that had sickened a four-year-old Georgia girl.

Among his mounting evidence of Lyme disease in the South, Clark had reported case studies of the disease among residents who had never left the area. In his own blood, he had found the DNA of *B. burgdorferi* and another exotic species, *Borrelia andersonii*, after he was bitten by a lone star tick on a trip to collect ticks in Fayetteville, Georgia. That field research left him with a years-long case of Lyme disease—or what he might call Southern Lyme-like illness but certainly not STARI. That vague descriptor, he told me, minimizes and leads to missed and improper diagnosis of the disease.

Denying an Epidemic

It was against this backdrop that Rudenko and Clark joined forces in 2013 with a legend in the science of southern ticks named James Oliver Jr. Oliver, like Clark, had challenged prevailing wisdom on potential Lyme disease in the South, first identifying blacklegged ticks in Georgia in 1993 that had been mistaken for another species. Their project was one all too rare in the science of tick-borne disease: a study of people, their ongoing illness, and its possible cause. Together, they collected and analyzed blood and plasma samples from twenty-four patients from the South with a constellation of ongoing symptoms including "severe headache, nausea, muscle and joint pain, numbness and tingling sensations in extremities, neck pain, back pain, panic attacks, depression, dizziness, vision problems, sleep problems, and shortness of breath." Among them, 71 percent recalled a tick bite and half had gotten a rash. All had previously been treated for suspected Lyme disease, for which they all, nonetheless, had tested negative. Clearly, this study had three strikes against it from the get go, as far as deviating from mainstream dogma. The researchers were looking for Lyme disease in a verboten area. They were studying patients who were "seronegative"—they didn't pass standard two-tiered testing. Their subjects had posttreatment conditions that suggested chronic disease.

The team found something. In fact, it might be said they hit a small but significant bit of pay dirt. From the fluids of three patients, the team was able to culture Lyme disease spirochetes, a feat that fails about half the time even when a rash is present and infection is certain. In two Georgia residents, the scientists isolated *Borrelia burgdorferi*, the first time the spirochete had been cultured in residents from the South. In the third patient, from Florida, they found a strain of a rare species, *Borrelia bissettii*. Evidence of *B. bissettii* DNA had already been detected in human samples from the Czech Republic and California. This was the first time the organism itself had been cultured from a patient in North America. In all three cases, this was not spirochetal DNA, often dismissed as the dead remnants of infection. This was living bacteria.

The resulting article struck a nerve at the heart of the Lyme disease divide: "The successful cultivation [of spirochetes] from antibiotic-treated patients might suggest that active infection with persistent *Borrelia* may be the cause of recurrent symptoms and persistent disease," it stated. Beyond touching on the nuclear issue of Lyme chronicity, the article included a blunt assessment of diagnostics. "This study would never happen" if the patients had been tested "according to CDC guide-lines," Rudenko's team wrote. Rather, the patients would have been considered "inappropriate" for study because of where they lived, how they tested, and that they had previously been treated with supposedly curative doxycycline. Yet here were at least three patients, 12 percent of the group, with significant symptoms and active infection that was validated by the gold standard: live cultured organisms.

Using these findings, which were published in the journal *Clinical Microbiology and Infection*, Rudenko attempted to obtain funding for additional research. Her proposal was rejected with a cutting comment from what she described as a US-based reviewer, whose identity, as in all such reviews, was shielded from her: "I (and most other experts) are extremely skeptical of the publication on which this whole project is based. Certainly there should be evidence that the findings can be reproduced by others before good money is wasted pursuing this line of investigation which almost certainly will be fruitless. I suspect that the investigator is not responsible for fabricating data, but rather that the samples from the United States that were the basis for the publication were tampered with." This wasn't just skepticism; this was rejection on grounds that had little to do with the science.

"So, that is how it is," Rudenko wrote to me in an email in the winter of 2017. In my research for this book, I learned of similar experiences of Lyme disease researchers who try to tell another story. They told me they were not believed—that their science was rejected outright or ignored. The contentions of Rudenko's grant reviewer were perhaps the most blunt example I had yet come across. A cou-ple of days later, I shared the reviewer's comments with Rudenko's coauthor, Kerry Clark, in a conversation about Lyme disease in the South. He was appalled at the inference. "It shows you the depth of some people's fear of being proven wrong," he said. At the heart of the issue was whether Lyme disease was real in the South and—the larger implication—whether the tick-borne disease was not just a minor irri-tant there but a serious public health threat.

A cadre of residents of Florida, Georgia, South Carolina, and North Carolina believes the latter. They meet in support groups in church basements, set up fundraisers to support Clark's research, post stories of misdiagnosis on websites, and write letters to health officials and legis-lators. This surely does not an epidemic make. But these agitators, like their comrades in other states and countries around the world, have been trying to tell us something, as is the science behind them. They have instead been told that science and medicine have it covered on Lyme disease in the South, and that it has been found to be rare.

Nonetheless, consider this. North Carolina has one hundred coun-ties, just five of which had been declared "endemic" for Lyme disease by 2017, a designation given by health authorities when at least two cases are confirmed in each. But say a patient in North Carolina goes to the doctor with what looks like an erythema migrans rash. In an endemic county, a physician will call this Lyme disease. Just over the border—a line that ticks and the mammals that carry them frequently ignore—the diagnosis will be STARI. No tests are needed in the endemic county to confirm Lyme disease; tests are required to confirm Lyme in the non-endemic county. But if the test comes back positive in the nonendemic county, it will frequently be interpreted by a doctor to be false. After all, if the county is not endemic, physicians have repeatedly been warned, Lyme disease is unlikely. Be suspicious of potentially false positives, the dogma goes. This scenario, played out many times as reported by patients, serves to reinforce a highly questionable status quo. It keeps reported Lyme cases low—and patients untreated.

"It is interesting to contemplate the dilemma created by this prac-tice," wrote Marcia E. Herman-Giddens, a member of the Tick-Borne Infections Council of North Carolina in 2012 in a letter to *Clinical Infectious Diseases* journal. "Without positives, there is no Lyme disease;

if there is a positive, it is likely false." A clear example, she wrote, of "circular thinking." Indeed, this kind of thinking pervades Lyme disease, reinforcing prevailing beliefs generally as well as in the South. As a press release on a study by the United States Geological Survey put it in January 2017: "Lyme disease is very uncommon in the South." The USGS study announced then, as reported in the previous chapter, had analyzed blacklegged ticks—the kind that carry Lyme disease—and found them highly averse to southern heat. They consequently stayed hidden in the brush and away from human beings. But lone star ticks, the study dutifully noted, are well acclimated to southern living and "readily seek hosts high up in the vegetation well above the leaf litter."

That's where Kerry Clark found them, took them in for testing, and teased out *Borrelia burgdorferi* from a few. It wasn't easy, likely because not many ticks were infected and the concentration of spirochetes was low. Clark's findings of Lyme disease in Florida and Georgia patients were so controversial that a team of government scientists tested more than a thousand lone star ticks in an attempt to replicate them. The study, published in 2015 and led by the US Army Public Health Command, found "weakly positive" indications of the organism in a small share of the ticks but concluded that *Borrelia* infection in lone star ticks was "not confirmed." So two major questions remain unanswered for Lyme disease in the South: Does the lone star deliver the infection? And if not, precisely what organism causes STARI?

Without doubt, *B. burgdorferi* barely survives in the more challenging interior of the lone star tick, with which it shares an uneasy coexistence. The pathogen, meantime, thrives aboard blacklegged ticks, the ones that are more apt to hide in the southern shade. But as the next chapter shows, people in states like North Carolina are exposed often to and bitten frequently by lone star ticks. This tick has a far different demeanor and personality than its blacklegged counterpart. It is fierce, if a tick can be called that, and it is plentiful. Herman-Giddens said she found one on her arm once after a brief walk from her door to her car. "They just seem to appear out of nowhere sometimes," she told me, giving her ten to twenty bites a year. Said Clark, "In some places they are so abundant and attack in such numbers, that they almost seem to be dropping from the trees." Indeed, it practically rains lone star ticks in the South. That may be all it takes.

CHAPTER 8:

Not Just Lyme

"Lyme disease is the tip of the iceberg. There are worse diseases coming down the pike."
— Durland Fish, entomologist at Yale School of Public Health

When a mile-wide tornado ripped through Moore, Oklahoma, in 2013, killing 25 people, a shaken lot of suddenly homeless dogs became subjects in a study of disease spread by ectoparasites, among them ticks. Wandering among the rubble and brush of a battered suburb of Oklahoma City, the canines were like the white flannel flags that researchers sweep over grasses and leaves in efforts to snag and count ticks in the wild. Everywhere the dogs went, ticks, ever alert to body heat and breath, reached out with scabrous forelegs to latch onto bits of whisker or fur, the next meal within grasp. The project of a doctoral student at the veterinary school at Oklahoma State University in Stillwater, the study tested the blood of the forty-four wandering pet dogs and sixty-one others relinquished to a shelter. Fully ninety-two of them, or 88 percent, were infected with at least one tick-borne disease. In all, antibodies to four tick-borne infections were detected in the five-score and five dogs. Ticks have been called "cesspools of infection," and these tornado-tossed pups and their surrendered brethren showed why.

As is the joy of scientific pursuit, the Oklahoma researchers discovered something else in their study of homeless canines. A single Oklahoma dog was the state's first documented case of infection with something called *Rickettsia amblyommii*, a novel pathogen, perhaps up and coming, in the Rocky Mountain spotted fever family and delivered, as ticks go, by a particularly nasty one that you met in the last chapter. The lone star tick, distinguished by the female's golden crest on her chestnut back, has abilities that a Lyme-toting blacklegged tick can only dream of. Entomologists commonly describe it as "aggressive." No, they aren't "hostile or belligerent," a spokesperson for the US Centers for Disease Control and Prevention responded when I asked for a definition. Rather, "these ticks can actively move toward a warm-blooded host when they detect its presence nearby." Whereas the blacklegged, aka *Ixodes scapularis*, tick moves one way—vertically up and down on grasses and twigs in search of a passing mammal meal—the lone star, or *Amblyomma americanum*, tick can crawl horizontally as well in its quest for a host. While the blacklegged tick in North America or castor bean tick in Europe passively waits for prey, the lone star hunts, stalks even, finding a farmer at work in Missouri, a camper setting up in North Carolina, a white-tailed deer in the tallgrass prairie of eastern Nebraska. One scientist told me it takes seconds for a lone star tick to climb up a pant leg, pass over a shirt, and find its way to the soft nape of the neck or the safety of the scalp.

The United States military with its sprawling bases and outdoor exercises has reason to be concerned about such things and has long kept an eye on resident ticks. At the Aberdeen Proving Ground in Maryland, for example, soldiers in training, who often creep belly to the ground through brush and grass, have encountered up to five hundred nymphal lone star ticks—per soldier, per hour. A nymph is in the second of the lone star tick's three life stages, when it has already fed once and, therefore, may have picked up any number of infections from its "host" animal, usually small mammals. That was not good for soldier health. Nor was what they found in the ticks.

In samples from thirty-three sites on the base's 72,500 acres on the Chesapeake Bay, Aberdeen researchers found "widespread distribution" of ticks infected with the bug that causes human monocytic ehrlichiosis, a disease that infects about 1,500 Americans annually and can be fatal if not treated promptly. By 2012, the Army researchers had concluded the lone star was no longer the "nuisance pest" it had long been regarded. Instead, it now carried three varieties of the pathogen that caused ehrlichiosis, and its role in two sometimes fatal diseases, Rocky Mountain spotted fever and tularemia, was "being re-examined."

To that point, Lyme disease from *Ixodes* ticks was known to be widespread in the military, reported at 120 military bases, from West Point in New York to Landstuhl in Germany and in every region of the United States except Hawaii. Cases topped 3,200 from 2001 to 2008, although the true number was likely much higher, using the CDC's tenfold undercounting yardstick. Soldiers had reported Lyme cases that led to discharge when they could no longer do the job. Now, the tick-borne menace was growing, in size and scope, and not only for the military. In Oklahoma, where the tornado-tossed pups were studied, historical records unearthed from libraries and government archives had shown lone star ticks living in eighteen counties. By 2015, researchers found the ticks in sixty-eight counties, 88 percent of the state's total. This then is the future of tick-borne disease: more ticks in more places with more pathogens.

In 2015, scientists from four US government agencies assembled data to predict that future, to model how far, in the latter half of the twenty-first century, the lone star tick might move, driven by the winds of climate change. First, they studied conditions in the places the tick currently lived. The model looked at how much snow there was in

October; the temperature at the wettest and driest times of year; the difference between high and low temperatures; the benchmarks of warmth that help farmers predict the emergence of insects and blooms; and last, the vapor pressure, a measure of humidity. For ticks, which desiccate and die in hot, dry conditions, the humidity factor was by far the most significant for lone star survival. Then the scientists plotted where conditions favorable to lone stars might emerge in a warmer United States, circa 2061 to 2080. Disturbingly, they found vast swaths of new territory for *A. americanum*, which had once been relegated to the southern United States, Central America, and northern South America.

In coming decades, they predicted, the lone star tick would follow the pioneers who moved north into the Great Plains and the Ohio River Valley and the Scandinavians who settled the cold Midwest two hundred years earlier. But the ticks would be driven by something other than frontier spirit; they would thrive amid more humidity and more warmth. And that, the scientists found, meant "considerable increases in the proportion of suitable habitat in Iowa, Illinois, Indiana, and Ohio. With greater warming, southern and central portions of South Dakota, Minnesota, Wisconsin, and Michigan are also likely to increase in suitability."

Studies often note that some places might become too warm to support ticks, but this should be slim comfort. East Texas and coastal areas of Florida and Alabama, lone-star research speculated, might become climatically unattractive to the lone star. But the scientists were far less certain about what kills ticks off than what propels them forward. "We don't feel we can say that ticks won't occur in areas in the future," one of the model's designers, Catherine Jarnevich of the US Geological Survey in Fort Collins, Colorado, told me. Other scientists agree this is not a zero-sum game, with ticks gained here but lost there. What is clear is that there will be more places for the lone star to live, thrive, and bite, a trend, the report said, "almost certainly driven by warming and the accompanying changes in humidity, particularly during spring and summer months." A year after that study, the US government's climate change assessment predicted that ticks loaded with Lyme disease and other pathogens "will show earlier seasonal activity and a generally northward expansion in response to increasing temperatures."

Tale of Two Continents

September Norman was a trim, blonde woman of fifty-four who liked hiking and lived in a leafy Tennessee city that juts into Old Hickory Lake known as Hendersonville. During a round of late-spring gardening in 2014, Norman had a chance meeting with a lone star tick, sustaining one of many bites she'd had over the years. This bite, however, would change her life in a strange and frightening way, one that would not become clear for another six weeks. That was when Norman awoke at 2 a.m.—in the middle of the woods on a camping trip—clawing furiously at an intense pain in her foot, in just the spot where she had last been bitten. She consumed copious amounts of water to flush her system, to no avail. Within a couple of hours, her tongue and lips had swelled; she was covered in hives, and her head looked to her husband like a giant red balloon. Ambulance attendants arrived as her throat was closing, and she was perhaps thirty minutes from suffocating. It would not become clear until she had another incident and was taken to Vanderbilt University Hospital in Nashville what had triggered the response. Both times she had eaten meat, first beef then pork. And she had been bitten previously by a lone star tick.

About a decade before, Scott Commins was an allergist at the Asthma and Allergic Diseases Center in the University of Virginia Health System when he and his colleagues started to see people like September Norman—patients with severe, sometimes life-threatening, allergic reactions. That's not so unusual in an allergy center except that, in these new cases, the response occurred after the patients had eaten mammal meat. Anaphylaxis, as the response is called, usually occurred almost immediately after exposure to a potential toxin, such as bee venom; these new reactions began three to eight hours later. Half a world away, Sheryl van Nunen, an allergist and medical school professor, had charted the same alarming trend at Royal North Shore Hospital in St Leonards, a city near Sydney on Australia's east coast. Van Nunen had seen her first case in 1987. As cases rose, she dutifully performed skin prick tests with raw meat and took extensive case histories. When a thread emerged linking meat consumption and the bite of a tick, the dynamic was so unusual that she remembers saying to herself, "This is indeed happening, and I am not imagining it." When she posited her theory, in an abstract for a conference in 2007, it broke the mold on the well-established immediate anaphylaxis reaction, not to mention the cause. The theory, she wrote in an email to me, had "most of my colleagues thinking at that time that I might have slipped my trolley wheels!" Today, knowing how ticks are masters at manipulating human immune response—that is, after all, what an allergic response is—she said it makes far more sense.

It would be Scott Commins and his colleague, Thomas Platt-Mills, who would publish a paper in 2009 in which they identified a carbohydrate in mammal meat called alpha-gal as the cause of the intense and life-threatening response. Then, in 2011, they and other scientists reported what van Nunen had suspected in Australia: "Tick bites are a cause, or possibly the only cause" of the allergy in the United States. Their conclusion marked the first report of "an ectoparasite giving rise to an important form of food allergy," and a validation of van Nunen's work. The eerie parallels between the work of these physician-researchers, so far apart in geography, speak volumes about the worldwide nature of tick-borne havoc and the schedule on which diseases and, in this case, an emergent allergy are making their unwelcome debuts. For Commins in the United States, the culprit was the lone star tick. For van Nunen in Australia, it was the Australian paralysis tick, *Ixodes holocyclus*.

When Commins described the first cases of so-called alpha-gal meat allergy in the United States in 2009, he was aware of twenty-four of them. By the time September Norman was bitten in Tennessee five years later, Commins had amassed 1,000 cases in Virginia alone, and he estimated there were 5,000 in the Southeastern United States in all. Cases were subsequently reported in France, Spain, Germany, South Korea, Sweden, Switzerland, Costa Rica, South Africa, and Japan, though the numbers were small. Many children were also affected. In a study published in 2013 in the journal *Pediatrics*, researchers reported the profiles of fifty-one children in Lynchburg, Virginia, aged four to seventeen, who had severe, unexplained allergies. Of the fifty-one, forty-five harbored antibodies to alpha-gal. More than 90 percent had been bitten by ticks in the previous year.

The question, for researchers, was, why now? It has long been the same question asked about the emergence of Lyme disease, and the answer was, oversimplified to be sure, that there were more ticks. In Australia, surging numbers of bandicoots, a mouse-like marsupial, and small mammals that support the life cycles of ticks were believed to have played a key role. In the American Southeast, more white-tailed deer enabled vastly more tick procreation and more clashes with people. This was undoubtedly part of it. So were the ways people interacted with nature and the rise of communities that prized small parks and forest patches, which provided habitats for those mammals to thrive. But the fuel for the tick-disease engine, of which alpha-gal syndrome is but one small working part, is climate change. As the US government report on climate change confidently predicted, ticks move and multiply in a warming world, affecting more places and people. "I see ticks every day, all day," said Andrew Lucas, an emergency department physician at Macksville District Hospital in New South Wales on Australia's east coast. If he sees thirty people in a shift, Lucas said, four will have come in for tick bites. "People get 200 ticks on them all the time." For Lucas, who I met in 2016 at a Philadelphia conference of doctors who treat tick-borne diseases, the threat was personal. His wife had suffered for two years from what he believed to be Lyme disease, an illness not officially recognized in his country and for which she could get only ancillary care.

The study that predicted widespread movement of the lone star tick suggested, optimistically, that the research had served a valuable purpose. It helped point to "areas where public health information campaigns could be initiated proactively and where field studies of tick ecology...might be conducted." In truth, we have little else but warning people and studying ticks, both of which have nonetheless been done in spare and inadequate measure. And both have failed to stop or slow the pandemic.

Because of that single tick, September Norman cannot have fish cooked on the same grill as steak, or any dairy products at all, so vicious is her allergy to that one mammal molecule. She and thousands like her, currently and yet to come, are left with allergies that could be set off by a bit of beef broth dropped into a soup or some bacon drippings used in a gravy, a tick-borne legacy that lasts a lifetime. The lone-star's ability to impart a lifelong allergic response is significant and unprecedented in the annals of immunology. But it is not the only new discovery in the lone-star arsenal and is part of a growing list of pathogens and problems that ticks, lone star and otherwise, can cause.

The Next Big Thing

On a farm in central North Carolina in 2010, a sixty-one-year-old man found a tick in his right armpit. Within a week, he developed body pain, headache, nausea, and a fever of 103 degrees Fahrenheit. The man, who ran a small beef cattle farm, removed the guilty arachnid, a beautifully preserved-in-alcohol, engorged, male lone star tick. The farmer, it seemed, was also a leading researcher of tick diseases, based at the College of Veterinary Medicine at North Carolina State University, Raleigh. In a study of his case published in 2011, Edward Breitschwerdt and his colleagues concluded they had documented, through molecular testing, the first infection from a lone star tick of Rocky Mountain spotted fever, a nasty illness normally delivered by the dog tick, *Dermacentor variabilis*. "If you don't treat for Rocky Mountain spotted fever by the fifth day of illness," a researcher at the US Centers for Disease Control, Kenneth Dahlgreen, told NPR Radio in 2015, "there's a really good chance you're going to die. And it's an ugly, ugly death, too."

Breitschwerdt recovered well. Moreover, his case helped explain a strange new trend in tick spread and disease. Since the late 1990s, health authorities had watched the number of spotted fever cases soar, from around 350 nationwide in 1993 to about 2,000 by 2010. On the upside, as cases rose, the share of people who died, namely the fatality rate alluded to by Dahlgreen, had dropped sharply—another indicator that something new was happening. So Dahlgreen and his colleagues at the CDC plotted where cases had grown and fatality rates had dropped. It was when they added one more variable—where the lone star tick was prevalent—that the picture became clear. "The expanding range of *A. americanum* is associated with changes in epidemiology" of Rocky Mountain spotted fever, the researchers reported in 2015. More lone star ticks in more places, they concluded, meant more Rocky Mountain spotted fever.

Breitschwerdt, to be sure, got there first. Though he contracted an infection that the CDC states "can be a severe or even fatal illness," he was nonetheless jubilant over his luck. That tick tucked in his armpit, he told me after a round of baling hay on his farm, was nothing less than a bit of "medical serendipity," the kind of thing science is sometimes built on. "If I was not a vector-borne disease researcher, I would not have saved the tick and perhaps we would still not know if the lone star tick could transmit RMSF to a dog or human," he said.

After thirty years of studying ticks, Breitschwerdt was less concerned about spotted fever—though it most certainly was a growing threat—than about another infection he believed would eclipse even Lyme disease in size, scope, and significance. It often accompanies Lyme as one of several leading coinfections that vastly complicate diagnosis, care, and recovery. But tagging it onto Lyme disease failed to accord this bug its proper due. Decades of study had convinced Breitschwerdt that this pathogen would stand alone in the pantheon of emerging human diseases: *Bartonella*.

Like Lyme disease, the infection was driven, though certainly not caused, by climate change. It was also supported by small and big mammal hosts like mice and deer. More troubling, however, were many other so-called "reservoirs" of infection and many other methods by which *Bartonella* was delivered to people—the bite of a tick, sand fly, flea, or louse (indeed, many children do not get through grade school without a brush with head lice)—and, more classically, through the scratch of an infected cat. An epidemic hiding in plain sight, as Breitschwerdt called it, the pathogen was not identified until 1990 and only after it had emerged in an AIDS patient seven years earlier.

By the early 2000s, Lyme disease specialists were finding *Bartonella* infection, or bartonellosis, in growing numbers of patients. That may not have been surprising, given the widespread prevalence of the pathogen in the environment and the many ways to contract it. A key question remained, however. Was it tick-borne? At least nineteen research studies from 1999 to 2016 had found the pathogen's DNA in ticks. Rates of ticks infected with *Bartonella* came in at 18 percent in France; 29 percent in China; 44 percent in Russia; and, in three American states, 35 percent in New Jersey, 13 percent in Indiana, and up to 19 percent in three California studies. Those results certainly pointed to ticks as a potential delivery method of *Bartonella*.

But tick-borne disease is nothing if not controversial. A 2016 review of the literature, by researchers in Frankfurt, Germany, found multiple studies that showed *Bartonella* in ticks in California, the Czech Republic, and France, with an average of 15 percent infected in Europe and 40 percent in Finland. The review also found ticks, dogs, and people with dual infections of *Bartonella* and Lyme disease together, suggesting tick involvement. But in an example of the power of one side of the Lyme divide, the review said the jury was still out, the issue "still controversially discussed." It referred three times each to two studies, all written in part by Gary Wormser, the lead author on the Infectious Diseases Society of America Lyme disease treatment guidelines. His influence is considerable, his name cited in hundreds of papers. "Clearly," the *Bartonella* review concluded, citing Wormser, "*Bartonella* DNA which was found in several tick species in multiple studies does not prove the presence of viable bacteria." By then, Ed Breitschwerdt, a veterinarian by training, had seen enough of the *Bartonella* pathogen to be convinced that tick transmission was likely, even if it wasn't the only method of delivery.

For Breitschwerdt, the threat of *Bartonella* is linked to its ubiquitous presence in nature: mice, deer, squirrels, foxes, ground hogs, rabbits, bats, even kangaroos, may harbor the pathogen. *Bartonella* has been found in red-winged blackbirds in North Carolina, in bottle-nosed dolphins—more often in stranded ones, suggesting debility—and in sea otters in Alaska and California, though not necessarily associated with illness. Most significantly, cats carry it, putting veterinary workers and cat owners at particular risk from scratches or bites from fleas. "*Bartonella* is as complicated as Lyme disease," Breitschwerdt said, "and due to the worldwide distribution and numerous vectors"—ticks, fleas, and so on—"may prove to be more important on a global basis." Many physicians who treat Lyme and other tick-borne diseases agree that *Bartonella* is a significant threat to human health.

A Cat Bathed, a Baby Lost

In 1998, a man and woman who were raised and married in Long Island, New York, gave birth to twins, a boy and a girl, after years of trying to have children. The father was a carpenter, the mother, a graphic designer. The mother would not know for another decade that the care she extended to a ferocious, feral, flea-infested cat, bathing it twenty years earlier when she was sixteen, would potentially cause her and her family endless heartache and be the stuff of groundbreaking research. The twins, conceived through in-vitro fertilization, survived the caesarean section birth, but the little girl died nine days later. The cause was given as a defect on the left side of the baby's heart that impeded blood flow, also known as "hypoplastic left heart syndrome." Long before the births, the couple had suffered many unexplained health problems, including weariness, headaches, urinary and genital pain, atypical forms of pneumonia, and bowel issues. It was after the surviving twin developed a succession of problems—severe colic, night sweats, and hyperactivity—that the woman sought out an unlikely researcher: a veterinarian in North Carolina named Ed Breitschwerdt. She wanted to be tested for *Bartonella* infection.

By the time Breitschwerdt enrolled the family in a North Carolina State University study in 2009, this much was clear: *Bartonella* had been found in the embryos and offspring of infected white-footed mice and cotton rats, a clear suggestion it could be passed in utero in people. Infected mice also had higher rates of fetal death and low birth weight. Infected cats had trouble maintaining pregnancies. The hormones of pregnancy had even been suspected of promoting the growth of the bacterium in mice, cats, and cows. The couple's loss and ongoing trauma were both a tragedy and an opportunity. Breitschwerdt could explore the implications of this bacterium for people. When he tested tissue from a cervical biopsy the woman had undergone in 1991, the scientist found *Bartonella*. When he tested liver and brain tissue from the baby who had died days after her birth, Breitschwerdt found *Bartonella.* The father was infected with the same *Bartonella* strain as the mother. *Bartonella* infection was also found in their surviving son.

Ten years after their baby had died, after a decade of problems with their surviving child and other failed fertility treatments, the couple at last had an explanation, thanks to Bertschweidt's research. The cause of their grief was not an inherited defect or the mere product of chance. It was not of unknown etiology, or cause. Very likely, a small bacterium floating in the environment, delivered by the claw of a cat or the bite of a flea or from some other insect—few of us will escape exposure to *Bartonella* though we may remain asymptomatic—was why this family had suffered so much. Yet no doctor had been able to tell them this until they turned to a veterinarian in North Carolina.

In April of 2011, after publication of that case, Breitschwerdt received an email from a veterinarian he had known as a vet school professor years before: "15 years ago, my cardiologist told me I was in an 'elite group' of individuals as pertaining to cardiovascular fitness," the message began. "Last week, we discovered I need a mitral, aortic and tricuspid valve replacement." The vet suffered endocarditis, a heart infection. Like the Long Island woman, he suspected it might be related to *Bartonella.* The veterinarian's death at sixty-seven, and the death of another veterinarian at sixty-three, led to an article by Breitschwerdt in 2015, entitled "Did *Bartonella henselae* contribute to the deaths of two veterinarians?"

Three years after the vet's death, Breitschwerdt would find that 28 percent of 114 veterinarians and their technicians were infected with *Bartonella.* He studied these and other high-risk people who had serious neurological and other problems, finding *Bartonella* in two veterinarians; a golf coach who had lived on a farm; a horse farmer; and a young woman who had been severely scratched by a cat and suffered disorientation, motor problems, and seizures. He found it in a fourteen-year-old boy who, after a tick bite, developed migraine headaches so severe he was hospitalized. In Israel, 3 percent of long-term *Bartonella* sufferers had "often severe and disabling" joint disease, one study found; a study of twenty-two endocarditis victims from four countries, six of whom died, found nine infected with known *Bartonella* species, thirteen with strains not yet identified.

As with Lyme disease, *Bartonella* testing is imperfect. Standard antibody tests are notoriously inexact—less than 50 percent of infected patients will correctly test positive early in the disease, even fewer later—and these look for just two species of the bug, of perhaps a dozen so far identified. "*Bartonella* serology is a mess," said Breitschwerdt, who has founded a lab to test for the pathogen, "and yes, it is a major diagnostic problem." That is one reason bartonellosis has failed to be recognized, to reach the "tipping point" Ed Breitschwerdt believes is long overdue.

One of Many

If *Bartonella* alone can lead to infant death, long-term neurological problems, psychiatric problems, fatal heart infections, seizures, and brain inflammation, consider what it might do when paired with any of a number of other infections that ticks carry. Then consider the diagnostic challenges of identifying, let alone treating, them. The patient advocacy group Lymedisease.org published the results of a survey in 2014 of 3,000 patients who had late-stage cases of Lyme disease. Among them, more than half had been diagnosed with Lyme and at least one other infection, and nearly a third had two or more. The survey only included patients diagnosed on the basis of the early Lyme disease rash or CDC-accepted laboratory tests. The list of infections was telling—a veritable menu of pathogens passed from ticks to people: 32 percent reported a coinfection with babesiosis, which causes a malaria-like illness; 28 percent had bartonellosis; and 15 percent reported ehrlichiosis. Also prevalent: Rocky Mountain spotted fever, 6 percent; anaplasmosis, 5 percent; and tularemia, 1 percent. (In addition, 15 percent had been diagnosed with mycoplasma, an infection that may or may not be tick-borne—there is scientific dispute—but that unquestionably seizes on compromised immune systems.) Recall the survey of 110 Canadian patients who had been diagnosed with Lyme disease; 36 percent said they were also diagnosed with babesiosis; 33 percent with bartonellosis. Three or more infections were reported by sixteen respondents. These nonrandom patient surveys surely overrepresent the general prevalence of coinfection and report illnesses for which diagnostic tests are sometimes uncertain. They are nonetheless indicative of illness in the sickest segment of Lyme patients, which is considerable.

Multiply-infected people should be no surprise, given reports of multiply infected ticks from virtually every country with *Borrelia*-laced arachnids. In some places, the rates are alarming. In 2016, 45 percent of *Ixodes* ticks from the French countryside were reported with at least one infection, and nearly half of those had two. In the Netherlands the same year, a third of *Borrelia*-infected ticks "carried another pathogen from a different genus," researchers reported in *PLOS Tropical Diseases*. In 2017, 67 percent of adult ticks in Suffolk County, New York, were positive for the Lyme pathogen, and 45 percent were found to have more than one infection overall. Of perhaps greater concern, a 2014 study suggested that two dangerous pathogens might fare better inside a tick than one. Nymphal ticks, the ones most apt to infect humans, were 83 percent more likely to carry both the Lyme disease and *Babesia* pathogens than chance alone would predict, researchers writing in *PLoS One* reported. The finding hints at some kind of mutually beneficial relationship between the pathogens, or with the small mammals that carry them, that isn't good news for people. Humans infected with both Lyme disease and babesiosis, which is caused by *Babesia*, "appear to have more intense, prolonged symptoms than those with LD alone," wrote researchers for the CDC's Epidemic Intelligence Service in 2006.

But while the literature is rich with reports of ticks carting multiple pathogens, there is a huge problem for the unlucky multitudes who actually contract them. These are the most difficult cases, presenting with infections unfamiliar to many practitioners and for which treatments are unstudied and regimens unclear. In the Netherlands, these patients cross over to Germany or Belgium. From Canada, they come into New York. In England, patients told me of flying to California or Washington, DC, to see physicians willing to treat them. In Australia, 800 patients sent their blood samples to a laboratory accredited in Germany for analysis. Five tick-borne pathogens were detected besides *Borrelia*: *Chlamydia pneumoniae*, in 116 patients; *Ehrlichia*, 63; *Rickettsia*, 13; *Bartonella*, 27; and *Babesia*, 7. Many Australian patients report they cannot get care for these diseases in their country, in part because these test findings are dismissed.

This is where the tragedy of unchecked Lyme disease and its often-related coinfections meets the reality of a system unable to deal with either, let alone both. In Lymedisease.org's survey of late-stage patients, half saw seven doctors, waited ten years for a Lyme diagnosis, and traveled more than fifty miles for treatment. It isn't just that doctors do not know what to do in complicated cases, although they don't. The bigger problem is they have been told there is no problem at all.

Consider a 2014 article in the *American Journal of Medicine* that reviewed the evidence on tick-borne coinfections. Beyond challenging bartonellosis as a tick-borne disease, Gary Wormser, the treatment guidelines author, and Paul Lantos, of Duke University, dismissed the contention that multiple tick infections can cause lasting problems. In this qualified, carefully sculpted sentence, they concluded: "The medical literature does not support the diagnosis of chronic, atypical tick-borne co-infections in patients with chronic, nonspecific illnesses."

Translation: If the studies haven't looked at it, if we can't find it through imperfect tests, if doctors do not know to test for it, and if your symptoms are all over the place, you don't have it. To be sure, the literature is spare on how to diagnose and treat tick-borne coinfections. Indeed, coinfections were not tested for or considered in the four treatment trials of long-suffering Lyme patients that have dictated short-course antibiotic treatment for nearly a generation. These small, incomplete, and flawed studies have been used, in article after article, to reject the reality of persisting Lyme disease just as a lack of evidence is used to dismiss long-standing coinfections.

Lorraine Johnson, the executive director of Lymedisease.org, is a lawyer with a master's degree in business and about forty peer-reviewed publications. Her survey results appeared in the open-access journal *PeerJ*, which has about half the all-important "impact factor" of the journal in which the Lantos-Wormser paper was published. She is nonetheless undeterred.

Johnson compares her approach to the Framingham Heart Study, which found a new way, starting in 1948, to look at a major cause of death in the United States. Researchers enrolled thousands of patients, studied their lifestyles, took detailed case histories, and tracked them over decades. Ultimately, the study drew significant conclusions on the causes of the American epidemic of cardiovascular disease. In that tradition, Johnson's surveys of advanced Lyme disease patients, which like Framingham are ongoing, have found more than 40 percent of late-stage patients suffering severe fatigue and sleep disturbance; nearly 40 percent with severe joint and muscle pain; and about 30 percent reporting severe cognitive impairment, depression, and other pain.

To be sure, Lymedisease.org's patient survey isn't the whole story; it is what Johnson called "observation trials." These involve, she said, "real clinical patients in a clinical setting receiving treatments that real treating doctors prescribe." Her *PeerJ* paper admitted the limitations of self-reported data from a motivated population, just as traditional Lyme studies have reported potential biases. But she and her colleagues did a key thing that the science of the static, prevailing, traditional dogma has not: They listened to and considered what patients said.

The Unknown and Undiscovered

In October of 2001, the Dutchess County Department of Health issued a warning to physicians and laboratory directors in the small, green county of New York, which is located about a ninety-minute drive from the northern reaches of Manhattan. "Public Health Bulletin," it said: "*BABESIOSIS.*" Two residents of the bucolic county, bordered by the Hudson River on the west and the state of Connecticut on the east, had shown symptoms of what the alert called a "potentially severe and sometimes fatal disease."

Babesiosis is in the same protozoan family of diseases as malaria and is characterized by drenching night sweats, chills, body pain, fatigue, and anemia. Like Lyme disease, the illness is caused by the bite of an infected blacklegged tick, sometimes sickening its unsuspecting victims with the double whammy of both diseases at once, an experience victims would not wish on their worst enemy. For decades before the bulletin was issued, babesiosis had been considered a coastal problem, with the first US case emerging in 1969 on Nantucket Island off the coast of Cape Cod, Massachusetts. It became so common there that it was called "Nantucket fever." Connecticut had seen its first case in 1988 in the southeastern part of the state. In New York around then, intermittent cases were largely limited to Suffolk County on the eastern half of Long Island, which juts into the Atlantic Ocean for a hundred miles from the edge of New York City to its tip at Montauk Point.

The bulletin, welcome though it was, was not news to Dr. Richard Horowitz, an internist who, since 1990, had specialized in complicated tick-borne cases in a practice in Hyde Park, New York, famous as the Hudson River home of President Franklin Delano Roosevelt. It wasn't long before Horowitz, a blue-eyed Buddhist whose life-mission was tick-disease care and cure, began to suspect there was more than Lyme disease invading the inland idylls of the Hudson Valley. So did his patient Jill Auerbach, the antitick crusader I wrote about in chapter 1, who had already suffered a crushing bout of Lyme disease. Then she was bitten again in 1994 and infected with this novel, debilitating parasite that was not supposed to have been anywhere near her backyard garden in Poughkeepsie.

This is the reality of tick-borne disease in a growth era. Government and medicine does not move nearly as fast as shrews, mice, and deer. While ticks feast and multiply on mammals, while they nimbly move widely on birds, while they flourish in milder winters and stay active in months when snow used to fall, the natural response of government is to resist suggestions that something has changed. Jill Auerbach had sought medical care and been told she had lupus, landing at last in Horowitz's office. No, he said, those vicious sweats and the crippling fatigue were caused by a new kid on the block, up and coming and dangerous, a tick-borne pathogen called *Babesia microti*.

That's when Auerbach, a savvy, indignant survivor—sick though she was—began a decades-long quest against ticks, personal and public, to limit the damage that was coming. She hatched a plan. She would get Horowitz together over dinner with Rick Ostfeld, a scientist at the Cary Institute of Ecosystem Studies in nearby Millbrook. Here were two men who were on their way to becoming world-class Lyme disease experts. The reason: they lived and worked in a beautiful but forbidding place, where Frederick Church landscapes were swathed in pathogen-packed ticks. By the end of the dinner in 1998, Ostfeld had agreed to pack up thirty tubes, each containing one adult tick in a solution of 70 percent ethanol, for delivery to a testing laboratory. Separately, Horowitz would send 192 blood specimens for analysis. The results showed *B. microti*

DNA in Ostfeld's ticks—though he was skeptical of what seemed an inflated infection rate—and in Horowitz's blood samples. "This report constitutes the first evidence of coinfection among *Ixodid* ticks with Babesiosis in upstate NY," Horowitz wrote in a conference abstract presented in April 1999.

In 1992, a twenty-five-year-old woman named Lia McCabe was earning her MBA and working for an investment banking company in New York City when things went terribly wrong. She became confused, forgetful, and immensely tired. She had migraines and would stumble when walking. She went to at least fifteen doctors and was told she had multiple sclerosis and chronic fatigue syndrome before she was ultimately diagnosed with Lyme disease. Unable to walk or care for herself, she recalls the day when she was twenty-seven and a hospital administrator said to her, "You have to decide what nursing home you want to go to." McCabe, who told her story at a New York State hearing on Lyme disease in November 2001, was unbelieving at the turn her life had taken. At one point, her mother came to her bedside, and McCabe could neither speak nor move. "I felt like I was in a coffin and couldn't get out," she said. Her mother walked out, thinking she was asleep.

Like many patients with advanced tick-borne disease, McCabe's life was both a struggle to get well and a battle to get care. Her insurer balked at paying for intravenous antibiotics, so her parents did. The drugs helped her improve but she backslid when she went off of them. Then in 1999, seven years after she became ill, she went to see Horowitz. He diagnosed her with babesiosis on top of Lyme. A month into treatment, she felt her legs again. "My drenching night sweats disappeared, my need for sleep decreased, and my energy level improved," she told the lawmakers. "After six years of being trapped in my wheelchair, I've packed it away and removed the ramp. I'm living a life I didn't think was possible." If she had participated in the Klempner study that decreed long-time antibiotics have no value she said, "I'd still be in a wheelchair today."

But it wasn't antibiotics alone that turned McCabe's life around; it was the diagnosis of babesiosis and treatment with both antibiotic and antimalarial drugs. Horowitz likes to say, definitively and often, "Babesiosis makes Lyme disease three times worse." Indeed, a 1996 study led by Peter Krause, a foremost *Babesia* researcher, found that people with both infections had more symptoms and longer illnesses than patients with either infection alone. Of great concern, about one in ten Lyme disease patients in southern New England, Krause found, were also infected with babesiosis in areas where local mammals harbored both infections. And that, Horowitz knew from the patients he saw, was a bit of geography that was growing. Three years before the babesiosis bulletin, he had reported the results of his own treatment study of 120 patients who had been unresponsive to antibiotic treatment for Lyme disease, or had relapsed after it, and who later tested positive for antibodies to *Babesia*. He found significant improvement after they were treated both with an antibiotic and antimalarial drug.

Jill Auerbach's effort to bring recognition to babesiosis involved researchers from institutions in New York and Rhode Island, led to the collection and testing of scores of ticks and blood samples, and prompted formation of a county tick force to pay for it. She wrote dozens of letters, made endless phone calls, and lobbied legislators and health officials. This was not government leading the way. This was government being led, as it is, with mixed success, in many places around the world. It was not until March of 2002 that New York State health officials issued a warning to doctors about babesiosis. That was a decade after Lia McCabe was infected and eight years after Jill Auerbach. It was also three years after Horowitz's conference presentation and his warning that babesiosis wasn't just coming. It was already there.

Poison Blood

With symptoms like air hunger, in which patients cannot seem to catch their breath, babesiosis can be a devastating illness. But its implications are as disturbing for people who are infected but who do not get sick. Their robust immune systems apparently keep the infection in check. But lacking any sign of illness, it does not stop them from the altruistic act of donating blood, which in turn goes to the sickest and least able to fight the infection.

In the United States, *Babesia*-tainted blood has been transfused into a forty-four-day-old baby with malformed lungs, an eleven-year-old boy on chemotherapy for a brain tumor, a fifty-four-year-old heart transplant recipient, and three premature infants. Those patients survived. But at least eight people in the United States have died, and twenty-five worldwide, after contracting transfusion-related babesiosis, the leading cause of transfusion-related microbial infection deaths in the nation. Among the deaths were two babies under twelve months, among nineteen infected. But these reported cases—162 in the United States through 2009—may be woefully misleading, representing what an article in the *Annals of Internal Medicine* in 2011 termed "a fraction of those that occurred." At the same time, tests and procedures to keep *Babesia* out of the blood supply and out of donated organs are still in the experimental stage. "Babesiosis is killing people, and it's contaminating the blood supply," Dr. Raymond Dattwyler, an author of the Lyme disease guidelines, told me. "You gotta have an appropriate way to screen for it," he said, something that, for now at least, is "cost prohibitive."

In the scheme of things, the babesiosis numbers may be low. But trends are important in tick-borne disease, and all the indicators point up: the number of reported cases, the number of places from which those cases emerge, and the prevalence of the organism in the wild. Lyme disease grew from a smattering of cases on the US East Coast in the 1980s to 38,000 nationwide in 2015, which represents a tenth of the true caseload. Will babesiosis follow suit? From 2001 to 2008, cases rose twentyfold in the Lower Hudson Valley, from 6 to 119, sickening patients with AIDs, cancer, and splenectomy, conditions that made them susceptible to the pathogen; one patient died. Cases in the United States rose from 900 in 2012 to 2,100 in 2015, a 133 percent increase. As with Lyme disease, a great many cases are thought to go unreported and undiagnosed.

Just as that public health warning about babesiosis in Dutchess County came nearly a decade after Lia McCabe's infection, other countries are playing catch-up in a race in which the organism leads. In China, the pathogen had been detected in ticks and mice in nine provinces and Taiwan. Researchers in 2014 were able to count twenty-seven human cases in five broadly spaced provinces, in the country's far north, south, east, and west. They included one babesiosis patient who, before taking ill, had received a transfusion of potentially tainted blood, according to an article in the journal *Parasites & Vectors.* "If this was the case, screening of blood donors in this region is urgently needed," the authors wrote. "Human babesiosis may have previously been overlooked in P.R. [the People's Republic of] China due to a lack of medical awareness and the limitation of clinical diagnostic methods."

China is not alone. *Babesia* strains have been detected in human blood from Germany, Switzerland, and France. The organism has been detected in ticks in the Netherlands, though human cases have not been found, yet. Cases have been diagnosed in Australia, Great Britain, and Austria, where the first human babesiosis case was reported in 2003 in a hunter who became ill two weeks after a tick bite. Five years later, 441 of 864 ticks, or half, were found to be infected with the pathogen. In Canada—where scientists have documented an unmistakable northward migration of *Ixodes* ticks—the first human babesiosis case was diagnosed in the summer of 2013 in a seven-year-old Manitoba boy weakened by congenital deformities.

By the early 2010s, researchers were trying to unravel the forces that led to ticks becoming infected with both the Lyme pathogen, *Borrelia burgdorferi*, and *Babesia microti* in the coastal Northeast. They tried putting uninfected ticks on mice that carried both infections—and discovered a new dynamic, reported in 2014 in *PLoS One*: "Coinfection in mice increases the frequency of *B. microti* infected ticks." In other words, ticks picked up the infection for *Babesia* more readily when mice were also infected with Lyme disease. Three of the researchers explained the dynamic in a review in *Trends in Parasitology* in 2016. On its own, *Babesia* is an organism of "low ecological fitness" that struggles to survive. But with *Borrelia*'s help, *Babesia* apparently could evade the mouse immune system, "removing one of several of the ecological bottlenecks" that would otherwise keep this bug in check and, rather, promoting its circulation in nature. "The Lyme agent somehow was helping increase the amount of *Babesia* in the mice," Krause, the babesiosis researcher, told me.

This phenomenon is translating into growing *Babesia* rates in ticks in areas with Lyme disease. In Lyme-rife Dutchess County in 2001, infection rates for *Babesia* in ticks were reported at 3 to 9 percent. By 2014, 13 percent were infected. Babesiosis had once been considered a coastal problem. But here in this place, a hundred miles from the Atlantic Ocean, the human rate of that tick-borne illness was then the highest in New York State, eclipsed the following year by Columbia County, just to the north. These are all harbingers, and there are many more.

Block Island is a magical, windswept place of cliffs and rocky meadows, nineteen miles from the nearest ferry at Port Judith on the shore of Rhode Island. In 1991, the island became a laboratory for the study of tick-borne disease, attractive for epidemiological study precisely because of its isolation as one of a string of islands off the Northeastern US coast that are vestiges of sediment left when the last great glaciers receded more than 10,000 years ago.

Block Island is part Welsh countryside, part rugged Irish cliffs, except that it does not have the unflappable rain. What this island does have, however, are residents with strikingly high rates of tick-borne disease: 516 cases per 100,000 for babesiosis and 1,677 for Lyme disease, according to the results of a ten-year study published in 2003. Researchers chose to study the island in 1991 because of its isolation and its population, both of people and ticks. When they were finished scouring medical records, Peter Krause and his team concluded, in an article in the *American Journal of Tropical Medicine* that a tenth of the population showed exposure to the *Babesia* organism and the risk of contracting babesiosis and Lyme disease was, as they put it, "intense." This is a word not often used in scientific studies. Moreover, the team had taken pains to be "inherently conservative" in computing the per capita rates, using only symptomatic cases of babesiosis. They also included only those who reported to the local medical center, effectively excluding the many cases of disease taken home in the blood of tourists, who visit by the thousands every summer season. "Babesiosis may be acutely debilitating," Krause and company noted, "and mortality rates of 5 percent have been reported among patients with babesiosis."

As with Lyme disease, *Babesia* does not respect borders. In Ireland, a fifty-eight-year-old farmer in Galway, in the west of the country, received a blood transfusion after having his spleen removed. A year later, he became ill with babesiosis and died after twelve days in a hospital. When his death was reported in 1989, there had been just six babesiosis cases, four fatal, in all of Europe. By 2011, about forty cases had been reported in Europe, still low. But something unusual was noted then. Two people, a woman, thirty-seven, and a man, thirty-five, in the Alsace region of northeastern France, were diagnosed with babesiosis. It was the first report of healthy people infected with the parasite, and it had occurred in a known hotspot for Lyme disease. Like Jill Auerbach and Lia McCabe, other French patients were likely told they were suffering some other ailment because the bug was unknown and doctors, unschooled. "In Europe," the authors of the French report warned, "babesiosis is probably underdiagnosed."

CHAPTER 9:
Childhood Lost

Kara: "Nothing more we can do."

In October of 2016, I stood in front of a painting at the Metropolitan Museum of Art in New York City called *Poppy Fields near Argenteuil*, a small masterpiece painted in 1875 by Claude Monet. Within an ornate gilt frame, I saw the idealized countryside of an Impressionist pioneer: a meadow flecked with poppies and lavender, bordered on the left by two slender green trees and lit gently by a sky more cloud than blue. In this field stood a child, said to be Monet's son, Jean, at the age of eight, legs obscured by the generous wild grasses of the French earth. I can no longer look at such pastoral loveliness without seeing what lurks within. I know what is there. Argenteuil is about eight miles from the heart of Paris, where twenty-first century parks have sprouted signs, necessary in a country with perhaps 30,000 Lyme disease cases a year. *Attention aux tiques*, they warn. *Les tiques sont des parasites qui se nourrissent de sang et qui peuvent transmettre des maladies à l'homme, notamment la maladie de Lyme.* Translated: "Beware of ticks. Ticks are parasites that feed on blood and can transmit diseases to humans, especially Lyme disease."

Kara Wilson grew up in a place with natural beauty that rivals Monet's most breathtaking landscape. Wilson's ancestors blazed the Oregon Trail with wagons and on horseback five generations before her, setting up a cattle ranch in Fossil, Oregon, a rural outpost on the Butte Creek east of the snow-capped Cascades. Nestled in the shadow of the brown and green foothills of the John Day River Basin, Fossil—population 473 in the 2010 census—has all the markings of American frontier: A broad main street lined with pickup trucks; a well-stocked mercantile, and a seventy-mile ride to the nearest chain supermarket. Life on the ranch, 9,000 acres family-owned, meant that Kara, born in 1976, began riding horseback early and often, first snugly linked to a parent's chest at the age of six months, then at two, on her own horse. The Wilsons would take the kids, each in their own saddle, to move the ranch's several hundred head of cattle on a big-sky afternoon, all of them returning dusty, achy, and exhausted at the end of the day.

Kara was a week shy of seven years old in May of 1983 when, after one such foray, she and her siblings—a brother, four, and a sister, eight—jostled for first dibs in the bathtub, and all wound up in there together. That's when Nancy, who had seen hundreds of ticks in her time, found one so deeply imbedded on the back of Kara's leg that she had trouble pulling it off. The tick was noted in a visit to the health center that night, but no connection was made to it when Kara got sick. Not when she developed a spring flu shortly after. Not when she was wracked by fevers, had colds that wouldn't quit, had pain that jumped from joint to joint. A photo from that time, taken against a white bed sheet, shows two stick-thin legs with knees swollen to the size of oranges. Doctors said the illness must be juvenile rheumatoid arthritis, but it came with an awful lot of symptoms—like raging nighttime fevers, vomiting, and rashes—that surely weren't typical. It made the Wilsons wonder. They did not know then that a scientist by the name of Willy Burgdorfer, for whom the infamous Lyme spirochete was named, was on his way to collecting 715 western blacklegged ticks in southwestern Oregon. Two percent would test positive for *Borrelia burgdorferi*, he reported in 1985. Not much was known about Lyme disease in Oregon then or in the dark years that followed as the couple searched for a diagnosis and a treatment that fit. It is a road that many parents of Lyme know well, and the journey, for too many, is not so different today.

In the months after Kara's tick bite, Phil and Nancy Wilson watched their rodeo-loving cowgirl shrivel from 42 to 29 pounds, where she stayed for seven years. She left school in second grade and did not return until the 10th. She was diagnosed with assorted other diseases including leukemia, for which her parents refused treatment and were threatened with charges of medical neglect. They stood fast, aware of the death of a little boy who had been given chemotherapy against his parents' wishes. In time, Kara lost her ability to walk and to see all but shadows. Her immune system was ravaged, making her prone, as is Lyme disease's wont, to constant infection. At one awful point, in March of 1988, Kara, eleven, contracted spinal meningitis and was comatose for ten days. Phil and Nancy were each taken aside by a nurse they knew and told there was nothing more to be done. "Honey, you've got to let her go," the nurse told Nancy. Kara nonetheless woke up, took 7-Up from a straw, and, because she could not speak, blew it into a doctor's face. However mystifying and joyous that moment was, it was not the end of Kara's trials. This is a doctor's summary of her condition six months later and more than five years since she had become ill, after she was taken by ambulance to a Hermiston, Oregon, hospital:

"PHYSICAL EXAMINATION: Reveals a girl of about 12 years old who looks not to be much larger than about 5 years of age. She is nonresponsive. Her jaws were clamped tight....Her eyes deviated, one to the right and one to the left. Pupils were small but they did not react to the light....CHEST: Thin. Ribs could easily be felt. BACK: Very thin. EXTREMITIES: So thin there is almost no muscle present. The joints are very knobby. It is difficult to tell whether they are just knobby because of the lack of tissue or whether they are actually swollen."

The doctor then wrote this: "IMPRESSION: Dehydration, possible aspiration. Rheumatoid arthritis, possible Lime's [sic] disease. Marked inanition," which means a kind of lifelessness wrought by lack of sustenance.

I tell Kara's story for several reasons. For all the anguish Kara's parents went through and all the damage she sustained, her story has an inspirational quality and a happy, if bittersweet, ending. Kara was so sick, unable to rise from bed, that the family removed the shoes from her younger brother's horse and brought it into the house, tail swishing, hooves clomping. Calves and baby goats were tramped through at birthing time, newly hatched chicks too, all to strengthen Kara's will, "to find that joy," she said, to enliven her spirit. Friends peeked in from the first-floor window, close enough to play with Kara but not to expose her to germs.

Children whose Lyme and tick-borne disease has dangerously progressed sometimes prevail against the bug. But the story of runaway childhood Lyme disease is not primarily about the solution, which in Kara's case—in most—is controversial and will not work for everyone. Kara's and the other stories I will tell here are about the search. For the parents of a child with late-stage Lyme, the quest for a diagnosis and then for care is draining, time-consuming, fraught with difficult decisions, and expensive. Undoubtedly, many children are lost along the way, to disability or worse, to parents who love but are less able or endowed than Kara's. This is not a journey for the faint of heart because in this one infection, unlike almost any other, institutional medicine is of little help. There is no accepted standard of care. There is, indeed, very little care.

In cases like Kara's, thirty years ago, as in those today, late-stage Lyme disease presents parents with a guaranteed and painful quandary. They have children who suffer debilitating, life-changing symptoms of a disease that mainstream medicine does not acknowledge exist. The accepted version of Lyme disease is as an acute, curable infection. What these children have, the tenets of the Infectious Diseases Society of America (ISDA) hold, is something else. But it is not Lyme disease.

This is the nub of the Lyme controversy, the one that has stymied research and hardened medical practice. Picture yourself with a sick child in the middle of one of the biggest controversies in medicine today. Then add in that doctors may belittle you and reject any suggestion that your very ill child may have Lyme and other tick-borne disease. Tests don't work so you will be sent from doctor to doctor for an answer that must be anything-but-Lyme. Further, health insurance will not pay for treatments that deviate from the standard Lyme protocol, even if those treatments are safe and standard for other ailments like acne or tuberculosis. And though long-term antibiotics are not ideal, they are all we have precisely because the pleas of parents and patients have been ignored. "I can't believe this is happening," a mother from Thunder Bay, Ontario, Jennifer Bourgeois, told me as she related her emotional quest for care, including fourteen-hour car trips to a US doctor, her son, Brody, thirteen, moaning and prostrate in the back seat. "Nobody deserves to have disrespect and be shunned like my baby was."

In the Netherlands, an advocacy agency called the Interest Group for Intensive Child Care looked into growing complaints by parents who had been investigated by child welfare authorities. The group was shocked to learn that 50 of 168 reports over an eight-month period from 2016 to 2017 were from parents of children with Lyme disease. The reports coincided with a change in reporting procedures in which schools and others were encouraged to report anything suspect, and many did. Some of the children with Lyme had been removed from homes; some parents had been accused of inflicting illness on their children or failing to follow medical advice they believed was wrong. The group studied each child's records and concluded that none of the suggestions of neglect or abuse had merit. What the parents had done was what parents do. They had believed their tick-bitten children remained sick after treatment, had kept them out of school, or had sought treatments from doctors, sometimes in Belgium or Germany, who practice a different style of Lyme disease care. "Only medical treatments which are according to Dutch standards are allowed," said Vera Hooglugt, who analyzed the parents' reports. "Children are so sick they cannot go to school. The school says there is nothing wrong with you. They cannot walk. They are lying down all the time." So, under the new liberal reporting policies, the schools alerted authorities, in what can be a life-changing event that permanently estranges parent and child. "They don't understand what it is for a child and a family when there is a report, a false report," Hooglugt told me. "What it does to a family and a child."

In Canada and the United States, parents like Kara's have similarly reported that they faced charges of medical neglect over their Lyme care choices. A sixteen-year-old girl told officials at a Pennsylvania hearing on Lyme disease in 2001: "My mother, an RN who took me all over to find a diagnosis…was actually accused of Munchausen's by proxy" – namely conjuring or causing her child's illness – "because she was trying to get her sick daughter well." As with the Dutch parents, the accusation had been linked so often to Lyme disease that in 2005 an article in the journal *Medical Hypothesis* suggested it amounted to "medical misogyny," a way in which doctors had minimized, even criminalized, the complaints of concerned women for their children. Other mothers told me of similar insinuations by doctors or fear of it; a Dutch woman who had sought antibiotics for her little boy declined to be named for fear it would alert authorities. A Connecticut mother, Kelli Avci, spent eighteen agonizing months under investigation by child welfare author-ities who questioned her son's prescribed antibiotics and his gluten-free diet before closing the case without explanation. "I can't even begin to tell you all I've gone through," she said. Such is the strange, contentious and highly politicized world of tick-borne disease.

Troy: 'No Such Thing as Chronic Lyme'

Alison Murphy got the news in a hospital room on March 31, 2016; she remembers the moment well. It was when she lost faith in a medical system she once thought functional.

Just six weeks before, her son, Troy, had been a wiry fifth-grader who played soccer, loved to be-bop dance, and stood out in a family of five boys for his silly, endearing sense of humor. Then, her blond, bespec-tacled boy of ten had a second bout in six weeks of flu, this one much worse than the last. Troy had headaches, fever, and lethargy so intense he could barely move. He began to drag his right foot. His legs and stom-ach hurt; he could not eat. On March 11th, his parents brought him to a local emergency room, where he underwent an abdominal ultrasound and an overall workup. Doctors found nothing wrong and released him. Things got worse. Troy lost fifteen of his seventy-five pounds. He was in excruciating pain, hypersensitive to touch, and regressing physically and emotionally. On March 24th, he was admitted to a hospital for dehy-dration, where he was described in his chart as "uncomfortable, consis-tently crying, reporting constant pain." After eight days as an inpatient there, numerous tests, and an ordeal that was torture to a bewildered, angry, and uncooperative Troy, doctors had an answer.

"Troy's presentation," his discharge papers concluded, "continues to be most consistent with diagnosis of pain amplification syndrome." In other words, Troy's symptoms were not physical but psychological. One doctor suggested the problem might be "schoolitis." And so on his release from the hospital, this withered little boy was prescribed two antidepressants and referred to a psychiatrist. He was told to get out of bed everyday at 9 a.m. and instructed to use breathing techniques for his pain. This happened to a child who between his ER visit and hospitalization had been diagnosed by a pediatrician with Lyme disease, a potential diagnosis that hospital doctors did not pursue. There is no evidence of tests for tick-borne illness in his discharge papers, which include only this reference: "Mom states he is currently being treated for Lyme disease…and started doxycycline last week." That was it. "They never wanted to hear another thing about it," Alison Murphy recalled in recounting her son's ordeal.

Kara Wilson was infected decades earlier in a place where and at a time when Lyme disease was barely known. Her long search is perhaps understandable. But Troy was infected in the American state of Connecticut, famous as the place where Lyme disease emerged in a coastal town forty years earlier. Drive an hour from Troy's home, and there's Yale University, the source of groundbreaking research on the pathogenesis of Lyme disease. Drive another hour, and you get to New York Medical College, where the world's supposed leaders in Lyme disease care are based. Yet Troy was given a psychiatric diagnosis and denied care for physical disease. It must, the prevailing wisdom goes, be something else.

This is not an isolated event; a little girl diagnosed with tick-borne infections named Fiona, nine, had a disturbingly similar experience to Troy's at about the same time and at the same hospital. "It was almost like they were laughing at our lab results," which, ordered by an outside Lyme practitioner, were positive for several tick-borne diseases, her mother, Laura Radmore, said. Here, at the very epicenter of the Lyme disease pandemic, tick-borne diseases and their many complications go unexplored and untreated. Consider what happens in countries around the world.

Sweden: A tick-bite at age five, a childhood filled with sickness, and, finally, depression led Alice Wallenius, nineteen, to a clinic in England, where she received the treatment no doctor at home would give her. She got better. The Netherlands: Childcare authorities twice threatened to take Erica Vrijmoet's three children, whose symptoms had baffled Dutch doctors but who were diagnosed with several tick-borne diseases by a Belgian physician. For years, her youngest daughter, who was six when taken ill, feared being taken from her home. Australia: The mother of an eight-year-old girl named Sophie, infected since she was two, told me, "We avoid hospital at all costs." Doctors routinely tell patients, she said, "It's all in your head." United States: A Massachusetts mother, Jean Derderian, in discussing her son, then fifteen, who was desperately ill with neurological Lyme disease, said, "We're so isolated. There's no hospitals. There's no care."

These are children whose childhoods are lost because of the simple bite of a tiny, common, insidious creature known in North America as a blacklegged tick; in Europe, the castor bean tick; in Russia and Northern China, the taiga tick, cousins all in the sweeping, and growing, *Ixodes* family. In the United States, children five to nine years old have the highest rates of hospitalization for Lyme disease. Boys that age have long had the highest rate and number of Lyme disease cases—nearly 20,000 in the United States from 2001 through 2015 and ten times that when underreporting is considered. The next biggest group was boys, like Troy Murphy, from ten- to fourteen-years-old.

Troy Murphy may be a canary in a coal mine, akin to the cases of gay men in the early 1980s who turned up with an unlikely cancer called Kaposi's sarcoma that opened the era of HIV. He may be a sentinel, and, to be sure, there are more like him. Unlike stories of AIDS, however, tales from the Lyme epidemic have changed little from the 1980s to 2010s. That is, with one very big exception. Early on, the issue was Lyme disease, a single infection that left a sliver of victims inexplicably disabled and in pain. Now, the larger problem is often the damage done by a constellation of pathogens that live in the bellies of ticks, a concept widely accepted in the scientific literature but disputed at the hospital door. Parents, consequently, must seek help elsewhere, or they must take what is offered. For these unrecognized, unacknowledged infections, medicine's menu is limited and sometimes harmful. It includes psychotropic medications, counseling—"cognitive behavioral therapy" was suggested to patients I met with in England—and psychiatric hospitalization. For more blatantly physical symptoms, there are steroids, minimal antibiotics, pain relievers, and a host of drugs used for arthritis, multiple sclerosis, and fibromyalgia. Many of these are ineffective and inappropriate because this approach fails to reach the underlying cause—an illness delivered in the bite of a tick.

After his first hospital stay, Troy continued on a downward spiral, with other unproductive trips to emergency rooms and hospitals. In December of 2016, eleven months after he got sick, I visited him at his home—barn-red clapboard with a white-columned porch and a fluttering flag—on an idyllic young cul-de-sac in a suburb of Hartford. There, a large living-kitchen area had morphed into a hospital room that featured get-well cards atop a white fireplace mantle and Christmas stockings hanging below. Troy, bedridden and attached to a feeding tube, was draped by a plaid wool blanket. His protruding feet splayed outward, a sign, along with eyes that jumped rather than tracked movement, of neurological disease. His thick blonde hair, once cut with military precision, now folded around his ears and shoulders. He sometimes moaned softly, sometimes grunted at offers of help from his mother, a slim, youthful woman with long blonde hair who had a way of reading Troy through long, probing glances, of knowing if he was in pain. Dr. Charles Ray Jones, a pediatric Lyme specialist from New Haven, Connecticut, was there too; he held Troy's hand and talked him through a gentle examination. Troy had distrusted doctors since his first hospitalization, when he had endured painful poking and prodding but had not been believed. Jones, eighty-seven years old and with a few thousand such cases behind him, had found a way around this.

A large man with black frame glasses and receding salt and pepper hair, Jones had decided to become a doctor in 1954 when, as a Boston University theology student, an old woman crippled by rheumatoid arthritis gently rebuffed his ministrations. "Please help me in a real way," he recalled her saying, cupping his hands in supplication as the woman had. Little did the woman know that Jones would become among the earliest physicians to find that antibiotics worked on a disease that, in early 1970s Connecticut, as yet had no name or known cause. He had observed, as doctors sometimes do today, that children with joint pain and fatigue got better when they were incidentally treated with antibiotics for strep throat. Jones' pediatric practice quickly filled with Lyme patients. For his penchant to prescribe long-term antibiotics, he had faced professional charges of medical negligence four times and incurred, he estimated, about a million dollars in lawyers' fees. When I spoke with him in the spring of 2017, he was facing an assessment by a neuropsychologist of his competence. On the wall of Jones' office is a quote from a friend at ministry school, Martin Luther King Jr.: "The ultimate measure of a man is not where he stands in moments of comfort and convenience, but where he stands at times of challenge and controversy." The quote is not about him, he said firmly.

When Troy was brought to an emergency room for dehydration in the fall of 2016, the Murphys tried, as many parents have told me they must, to avoid discussing the outside care for Lyme disease they had sought for Troy, to keep it, as Alison said, on the "down low." The couple knew where such admissions could lead. When she could no longer avoid the question, Alison told hospital staff of having sought Dr. Jones' care. That's when the presiding physician called Alison and William Murphy into the hallway. "There is no such thing as chronic Lyme," he told them in a way other patients have described, of doctors who are very sure they are right. "These supposed Lyme experts are a dime a dozen," the doctor said. "All they want is to take your money."

Charles Ray Jones, dressed in his signature jogging suit and sneakers and walking tentatively with a pronged cane, surely did not seem to be at Troy's bedside for the money. I had met him a month before at a conference in Philadelphia where, sitting in his wheelchair, he had told me about a desperately ill boy, Troy, who at that moment was in one of the best hospitals in southern New York, where doctors were giving mostly palliative care. Jones was beside himself with worry, his assistant told me then, incensed at the injustice of it. Like doctors at the first hospital in Connecticut, doctors at the New York hospital had refused to consider tick-borne disease in Troy's workup. By then, tests ordered by Jones showed markers for multiple, and treatable, tick-borne diseases. This is the new Lyme paradigm, the thing that makes Troy's case important. He had been exposed to the Lyme disease pathogen, followed by a damaging delay in treatment. Moreover, tests ordered by Jones showed the boy's blood littered with evidence of tick-borne pathogens, some suggestive, some definitive.

Troy tested positive for *Bartonella henselae,* a debilitating bacterium that can be transmitted by a tick, flea, or cat scratch; *Babesia duncani,* a parasite that causes malaria-like babesiosis; and *Borrelia hermsii,* also known as tick-borne relapsing fever. ("We don't have that in New York," a doctor at the hospital there had scoffed, Alison said.) Troy also had an "equivocal" reading for tularemia, which can cause wracking fevers and can even be a weapon of terrorism, likely explaining why it received more than half of all US funding for tick-borne diseases from 2006 to 2010. Beyond this, there was evidence of Lyme disease in Troy. The boy's blood had high antibody levels on one test for the Lyme spirochete and what Jones eyed as highly specific Lyme markers on three other tests, namely the "bands" on the second-tier test that I discussed earlier. Two physicians who had seen Troy before his second hospitalization had considered the antibody readings positive for Lyme disease, his records showed, although the findings, from licensed, certified laboratories, would not be ruled that way by the US Centers for Disease Control and Prevention. Too few bands.

Therein lies the Lyme disease quandary. Two-tiered testing is unreliable, missing a majority of positive cases early and an all-too significant share later, this in best-case studies. Yet alternative Lyme tests to which the sick must resort are not considered good enough. Then there is the increasingly common complication of other tick-borne infections. These, too, have been painted with the same brush as Lyme disease: exaggerated, unproven, and—for doctors who value their medical licenses—risky to treat. "Once you mention tick-borne illness," a Manhattan neurologist who had treated Troy, Dr. Elena Frid, told me, "all bets are off."

Joseph: Opportunity Missed

Diane and Ben Elone were proud, doting parents. They had two boys who were honing talents and pursuing dreams in a hardscrabble city on the Hudson River north of New York, where being a young black male was a challenge all in itself. The boys, a year apart, attended a high school in which 81 percent of students were impoverished and 39 percent failed to graduate. No matter. Diane, who worked in medical billing, and Ben, a thirty-year civil servant, were determined that would not be their children's fate. Emmanuel would go on to Georgetown and study in Europe. Joseph would sign up for a science program in which he caught, weighed, and released slithering eels at a local stream; he would attend a "summer scholars" program at Vassar College, a mile from his modest Cape Cod home on a tree-lined street in Poughkeepsie, New York. Usually reserved and quiet, Joe was jubilant after the program wrapped up, heading to a local hamburger joint where he and some friends spent several hours reliving the experience.

By his junior year, Joseph, seventeen, had found a calling of sorts. A boy who was often photographed outdoors—knapsack and black-frame glasses on, the majestic Hudson Valley unfolding behind him—Joseph wanted to protect the environment. That led to two weeks at ecology camp just after school let out in June 2013. He came back on July 5th more committed than ever. Then the problems began. Joseph developed what was initially seen as a "viral syndrome"—a sore throat, cough, and fever. He was tired and had body aches. Twice he went to the pediatrician. "They checked him for strep throat. He didn't have strep throat," Ben Elone said later, sitting at his kitchen table, Emmanuel by his side. "They checked him for Lyme disease; he didn't have that…The doctor figured that the coughing would stop. He just needed to drink a lot and take a lot of rest."

The clues, to be sure, were there. Just before he got sick, Joseph Elone had spent two weeks, largely outdoors, in Rhode Island, which then ranked seventh among states for Lyme disease in America, rising to fourth the following year. He lived in a county with a Lyme disease rate in 2013 that was four times that of sky-high Rhode Island's. His symptoms had appeared in prime tick season, when summer flus should be viewed suspiciously. Some of his symptoms—the malaise, the fever, the achiness—suggested Lyme disease.

But Joseph had no rash, at least not one that was seen against his dark skin, and his other symptoms suggested an upper respiratory tract infection. Moreover, his test had come back negative, falsely so as it does in more than half of early cases. The doctor merely followed the Lyme disease guidelines of the Infectious Diseases Society of America (IDSA).

At three different points, as I reported in chapter 6, these influential directives advise physicians not to treat potential Lyme disease in people without a rash who do not test positive. Symptoms simply "are too nonspecific to warrant a purely clinical diagnosis," the guidelines say, in cases involving neurologic, arthritic, and cardiac symptoms. When

I asked if the CDC concurred, a spokesperson referred me to this sentence in the IDSA guidelines: "Erythema migrans is the only manifestation of Lyme disease in the United States that is sufficiently distinctive to allow clinical diagnosis in the absence of laboratory confirmation." No rash, no positive test, no diagnosis. And so, for Joseph Elone, who lacked the tell-tale rash, no antibiotics were prescribed.

Twelve days after his last doctor's visit, Joseph needed cough drops—the doctor had said to try some—so his mother took him to a nearby pharmacy. His symptoms had persisted; his records state "he had developed diarrhea and lightheadedness and was reported to have photophobia," or sensitivity to light. The Elones were planning a third doctor's visit. Instead, just after returning home, Diane Elone looked out onto her small front lawn, with a large tree on each side of a walkway to the front door, and saw her youngest son lying face up. He wasn't breathing. Emergency medical workers resuscitated him there. In the ambulance on the way to a hospital, Joseph's heart stopped twice, and twice he was brought back. At the hospital, his pupils reacted to light, a good sign, but his other indicators were not good. His kidney and liver function were abnormal. His white blood cell count was high. A nurse suggested, not unusual for Poughkeepsie, that this young black man, unconscious and breathing shallowly, "must have taken something." But his urine was clean. What Joseph had in his body were Lyme disease spirochetes. They were in his lungs, in his brain, and, most significantly, in his heart.

In a helicopter on the way to a tertiary-care hospital several hours after his collapse, Joseph's heart began beating wildly and attendants used electrical currents in futile attempts to reset it. He arrived at the hospital unresponsive, on a ventilator and with a pulse that went from intermittent to none at all. He died a little after midnight on August 5, 2013. A few days later, Joseph Elone's second test for Lyme disease came back. It was positive.

In 1997, an article in the *Archives of Internal Medicine* reported on fourteen cases of a Lyme-like rash in children at a summer camp in central North Carolina; eight of the campers had recalled tick bites. Near the end of the article the researchers reported, in a reference so passing that it appeared in parentheses, that one of the fourteen campers had died, from tick-borne ehrlichiosis. The victim was twelve years old. In 2017, a plethora of ticks were discovered at a day camp in Lake Forest, Illinois, located in a wooded ravine along Lake Michigan. Officials thought it wise to move the camp to a suburban park. Imagine how many other camps in tick-riddled natural areas chose to operate in their usual fashion, with a few perhaps, many hopefully, doing tick checks, educating campers on tick avoidance, and aggressively responding to reports of achiness, fevers, and odd rashes. A college senior who helped with my book contracted Lyme disease a decade ago at one such camp, where her symptoms were not quickly recognized. It changed her life.

After Joseph's death, I studied records of 1.2 million deaths in New York over a thirteen-year period; I found that just nine had been attributed to Lyme disease even though regions of the state have among the world's highest rates of the illness. When I asked the IDSA to comment on my findings, a spokesperson pointed to a study that suggested that fewer, not more, people had died from Lyme disease than were being reported. The study, authored by researchers at the Centers for Disease Control, looked at 114 US deaths that had been officially attributed to Lyme disease from 2009 to 2013, of people aged nineteen to ninety-nine. Just one death, it concluded, was "consistent with clinical manifestations of Lyme disease."

In death as in life, late-stage Lyme disease is a controversial diagnosis, one that is disputed and discounted in ways that are not designed to ferret out truth. The CDC study looked merely at the coding of death certificates—some were incomplete, others deviated from procedure—and not at actual medical histories. These surely would have added insight. The study results were published as a "brief report" in the IDSA's *Clinical Infectious Diseases* journal, indicative of the effort that went into it. Beyond this, pathologists have not been encouraged to look for evidence of the tick-borne toll in postmortem studies that could shed light on disease progression. No registry exists to investigate suspect tick-borne deaths like the young camper's, or like three other North Carolina children and a two-year-old Kentucky boy, all who died from tick-borne Rocky Mountain spotted fever, their deaths only incidentally reported in the literature. Beyond sudden cardiac deaths like Joseph's, there are Lyme deaths from entrenched, systemic infection and from suicide, among them a thirty-three-year-old woman who stepped in front of a train in Wisconsin in 2014 after every avenue was closed to her. We simply do not know how many die from Lyme and tick-borne disease because no systematic effort has been made to look.

Kara: Search for Cure

Kara Wilson missed eight years of school to Lyme disease, and instead was tutored, read to, and engaged in other ways at home. Each year her parents dutifully did her hair—long and curly some years, straight and short in others—and dressed her up to take school-style portraits, the kind that document childhood. Kara's cheeks are full and rosy in the second grade. Over the years, the teeth grow but the facial structure seems locked in place. Her neck is lined; her eyes hollow. "You see a healthy, vibrant six-year-old," she told me, "and you see the decline into death."

In their search for a cure, the Wilsons took Kara to eleven pediatricians, assorted other specialists, and a hospital in California. They fled to a natural therapy clinic in La Gloria, Mexico, one step ahead of child protective services. It was at an integrative practice in Las Vegas, Nevada, where she arrived barely cognizant and in diapers, that Kara was diagnosed with Lyme disease in 1989. She was thirteen, and had been sick for six years. From there, the couple exhausted the possibilities of antibiotics for their daughter. She received electro-magnetic therapy, acupuncture, and homeopathic treatments. Some of those treatments, on both sides of the US border seemed to help, at times startlingly so, but they did not cure.

Then in 1992, the Wilsons signed Kara up for what became more than 150 sessions in a hyperbaric oxygen chamber, twice a day, six days a week for three months. The treatment, accepted today for control of serious infections and stubborn wounds but not for Lyme disease, delivers pure oxygen in a pressurized chamber, enriching the blood and prompting the killing of pathogens. Kara was part of a Texas A&M University study, designed by a researcher named William Fife to attack *Borrelia burgdorferi*. All ninety-one enrolled patients were positive for Lyme disease (though possibly not by CDC standards), and all had failed to get well with antibiotics. In 1998, after six years of trials, Fife reported that more than 80 percent of his patients improved significantly or were symptom-free. For Kara, it was the end of a long road. Then sixteen, her body may not have been restored, but her health was beginning to return. And so was her potential.

Shortly after her treatments, Kara, still unable to walk, went with her family to a rodeo. She suddenly realized, her fused fingers, calcified ankle, and weakened bones notwithstanding, I can ride. She bought a saddle on the spot, was tied to a family horse and off she went. "I ripped the Band-Aid off," she said, "and got back into it." Two years later, at eighteen, she graduated as valedictorian of her high school class, having clawed her way back from infirmity. She still rides horseback almost daily, often taking visitors onto the high desert range. Kara Wilson is nothing if not a survivor.

Bound to a wheelchair that does not much limit her, Kara told me her story in the midst of a roaring January snowstorm that had her worried for the ranch's four hundred head of cows. Deep snow is treacherous to cattle, every one of which needs forty acres to survive. Kara lives on and loves the ranch as her ancestors did; she runs the Wilson bed and breakfast where, over the years, she has told her story to thousands of guests. In 1999, Kara graduated from Texas A&M, a popular and beloved student. In 2000, she went down the aisle, in a wedding under a tent, on a bay roan quarter horse named Dynamite, her father on another horse beside her, and married a student she had met when she needed help getting on and off her horse. In attendance were 650 guests, including five well-meaning people who had reported the Wilsons for child neglect. In 2002, Kara and her husband, Brian, adopted a boy and a girl from Russia, where Muscovites in Red Square took pity on Kara in her wheelchair and dropped rubles in her lap.

Kara Wilson Anglin won her battle to survive, though in a changed and damaged body, in a way that Lyme disease patients occasionally do. Through the advantages of her birth, the fierce determination of a mother who rejected any hint of surrender, and a set of parents who fought off the sheriff and mainstream medicine to find what worked for their daughter.

There are infectious disease physicians, those who wrote the prevailing Lyme disease guidelines and those who accept them, who will read her story and be doubtful. They will say it was not Lyme disease that laid her so low. Even if it did, they will say, there's no proof that the treatment she received works. Perhaps. The problem is we have not studied what does work. There has not been a US clinical trial for Lyme disease care published since 2008, and the four done to that point were limited to three months of antibiotics. They did not consider the role of other pathogens carried by ticks. They did not try anything but two selected antibiotics, while Lyme specialists who deviate from the IDSA guidelines use a wide variety, sometimes in combination. The National Institutes of Health has spent an average $24 million a year since 2012 on Lyme disease funding, with nearly 400,000 cases in 2015, and, by contrast, $40 million on mosquito-borne West Nile Virus, with 2,000 cases. That's a woeful picture of underfunding. Even the CDC and

IDSA acknowledge that some people—from 10 to 20 percent, though figures vary—do not get well after treatment for Lyme disease. It is time to recognize a problem exists, and find out what does work.

Troy: A Disease of the Brain

"He's not walking. He can't sit up. He can't hold his head up. He acts like he's autistic or something."

—Alison Murphy, on her son Troy, 11.

That Lyme disease sometimes infects the central nervous system and brains of children, and adults, is not in doubt. Starting in the early 1990s, Brian Fallon and researchers at Columbia University had shown that children with long-standing Lyme disease suffered "significantly more cognitive and psychiatric disturbances," including behavioral, anxiety, and mood disorders. Tourette's syndrome, depression, bipolar disorder, attention deficit disorder, and dementia have all been associated with Lyme disease—and sometimes have improved with treatment. The physical manifestations of neurologic Lyme disease are equally varied: headaches, numbness and tingling, facial palsy, dizziness, unsteady gait, hypersensitivity to stimulation, and fatigue. This is a disease that affects both physical and mental health.

Within months of his twin bouts of flu at age ten, Troy Murphy had regressed in every way possible. He could no longer walk. He wasn't speaking. He could not eat. Neurologist Elena Frid looks at children like Troy, who was her patient, and sees brains on fire from tick-borne disease. The mechanics of this dynamic aren't fully clear, and cases like Troy's are not common. But the effects of late-stage Lyme disease have been documented in living color brain imaging called Single Photon Emission Computerized Tomography, or SPECT, scans. In a study of 183 patients with long-standing Lyme disease, published in the journal *Clinical Nuclear Medicine* in 2012, 75 percent had "discernible abnormalities" in the frontal, temporal, and parietal lobes of the brain. Here was something rare: what three physicians led by Sam Donta of Boston University Medical Center called "objective evidence" of damage to the brain in late Lyme patients.

Columbia researchers, similarly, used SPECT scans in 2009 to show how white matter in the brains of eleven Lyme patients had been starved of blood and therefore oxygen, fogging memory and sapping the ability to do everyday tasks. Ben Luft is a physician-researcher who was an author of the IDSA's Lyme guidelines in 2000, but who long ago broke ranks with IDSA over its dogmatic view of Lyme care. Fallon's Columbia team, he told me, unprompted, showed "a whole host of different abnormalities" that, nonetheless, traditionalists have ignored. "It's current, it's contemporary, it's cutting-edge," he said. "It hasn't been embraced." In a world in which IDSA tenets rule, doctors do not acknowledge symptoms like Troy's as physiological. Instead, he had schoolitis. Donta, similarly, comes from the other side of the Lyme divide. Also on the 2000 IDSA panel, he refused to go along with the finished guidelines in the belief they did not accurately reflect late-stage Lyme disease. As a consequence, his subsequent research, while peer reviewed, is dismissed by the IDSA side.

In 2016, a German researcher named Rick Dersch looked at all the studies—six of them—on treatment of children whose central nervous systems and brains had been infected with Lyme disease. The findings of the various publications were not informative. Studies were old, few, hampered by research biases, and confined to short treatment regimens. "Data is scarce and with limited quality," Dersch and his team reported in the journal *BMC Neurology*. Of great concern, the existing research was limited to treatment in the early stages of neurologic Lyme, when it is easiest to resolve. Tell us something we don't already know. As for cases like Troy's, there are no treatment studies to guide and inform physicians, and moreover no incentive to perform them. Both funding and urgency are in short supply. Yet the potential effect and toll of tick-borne disease on the brain is only growing.

In early November of 2016, a five-month-old boy ran a fever and began to vomit. The right side of his face took to twitching, and seizures set in. The boy, like Troy from Connecticut, had been bitten by a tick, this one armed with Powassan virus, a deadly pathogen transmitted in as little as fifteen minutes of attachment. Five of fourteen victims hospitalized in New York died within months, according to 2013 study in the journal *Emerging Infectious Diseases*, their health "severely impaired" by the infection. The boy fared better. Hospitalized and treated, he was, by ten months, meeting the milestones of infancy—babbling, crawling, walking with help. But there were indications that the virus had left its mark, as it does in half of cases. Brain scans showed he suffered a loss of brain tissue, called encephalomalacia. A form of calcification was also seen in a part of the baby's brain called the basal ganglia, a change usually detected only in scans of the elderly.

Discovered in 1958 after the death of a five-year-old Canadian boy, Powassan has been growing in recent years, showing up in states beyond the US Northeast, like Tennessee and Minnesota, and at unexpected times of year. The Connecticut boy had been bitten in the month of November, an outgrowth of increasingly balmy autumn weather, his case a harbinger. By 2013, the virus was reported in mammals in Alaska and Siberia, what a paper in *Emerging Infectious Diseases* called "rapidly changing Arctic environments." As blacklegged ticks spread on the heels of a warming climate and other environmental changes, so will Powassan, a Yale University expert on tick-borne disease, Durland Fish, predicted in 2015. "Until a few decades ago, it was only transmitted by a tick species that does not commonly bite humans," Fish said in an interview with Yale News. No more. "This recent change in the ecology of Powassan virus has caused concern within the public health community." In 2017, Powassan was reported in up to 5 percent of blacklegged ticks in Wisconsin and New York, while Connecticut deer with antibodies to the virus rose precipitously, from one in four in 1990 to nine in ten in 2009. Like that other new bug, *Borrelia miyamotoi*, Powassan can be passed from mother tick to her thousands of babies. That makes larvae infectious from their first blood meal and potentially spreads the pathogen to the mammals—mostly mice but maybe people—they tap first.

For better or worse, the growing number of Powassan cases, and a spate of them in the spring and summer of 2017, has helped to focus public attention on the risk of tick-borne illness. Here was an infection in which deaths and disability could be measured and seen, the line from bite to illness quick and distinct. Eighty miles from me, in the county in which my daughter and her three small children live, an unprecedented three Powassan cases, including one death, were reported that summer.

While Powassan's toll may be startlingly dramatic, the measure of damage by Lyme disease is to Lorraine Johnson crystal clear. It is documented in a set of photographs—her before and after brain scans—that are testament to six lost years of her life. A highly successful attorney who had overseen an entertainment company in California, Johnson in 1996 became bedridden; struggled to breathe; lost her ability to reason and think; and endured numbness, tingling, and arthritic pain. Doctors did not consider tick-borne disease, and though an avid hiker who had encountered ticks and had had a bull's-eye rash, neither did Johnson. She was diagnosed with depression, and, because nothing worked, cycled through more than fifty different rounds of psychiatric medications over the years.

The twin pictures, published in the journal *Psychiatric Times*, explain why her youngsters, five and seven years old at the outset, lost their mother for much of their childhood. This is another toll of tick-borne disease on children. In the left picture of Johnson's brain, as if the lights were out, is a dark red sphere with ugly blotches of blue and black around the eye sockets and temples. On the right is the same brain with the sun out. It is lit up in shades of orange and yellow with a hint of blue in the front. The first picture is Johnson's scan from 2002, when she was diagnosed with Lyme disease. The second, in 2004, is after she had been treated for two years with intravenous and oral doses of antibiotics for Lyme disease and several other tick-borne infections. "The depression never returned," the article stated in 2007. Despite some lingering arthritis and such, "she has mostly returned to her active lifestyle." When I spoke to her a decade after that article, Johnson agreed.

Alison Murphy, meantime, searches for signs of improvement in Troy who, when I visited him in late 2016, was early on the path taken by Johnson. Alison is hopeful. Troy cracks a joke sometimes. He is in much less pain. He drank some chocolate milk. Still, he does not walk. He does not lift himself from the bed in the living room. Maybe more time is needed. Maybe too much damage was done. The Murphys wait and wonder.

Among their concerns is how to pay a $487,000 bill for his seven-week stay at the New York hospital. The irony is that while he received little attention to his tick-borne maladies there, aside from about thirty days of sporadic intravenous antibiotics, Troy was given a diagnosis that shocked Alison Murphy. There in his papers, listed as Discharge Problem 2, after "loss of appetite," was the issue doctors assiduously avoided discussing or acknowledging: "chronic Lyme disease." The hospital nonetheless passed the buck, advising a follow-up visit with an infectious disease physician. "Will not send any prescriptions for the antibiotics…for Problem 2," the discharge document stated. That sentence exquisitely captures the radioactive quality of late Lyme disease in America and many countries. Let someone else handle it.

Joseph: Lesson Learned?

Joseph Elone's Lyme disease was missed under a set of care guidelines that require positive readings on each of two tests that its authors have defended so vigorously that no other diagnostic has been able to take its place. The trust-the-test mentality that permeates Lyme disease care, even in the face of risk factors and sickness, cannot be tolerated. Joseph's cause of death was Lyme carditis, a condition often referred to, on the one hand, as rare, involving one in a hundred reported cases, and on the other, as treatable. For Joseph, *Borrelia burgdorferi* bacteria impeded the flow of electrical impulses that controlled the beating of his heart. In medical parlance, the infection led to heart block. And it killed him.

In 2009, researchers at Children's Hospital in Boston studied 209 children with early disseminated Lyme disease—generally meaning the bacteria has moved from the site of the bite—and found that 33, 16 percent, had spirochetes in their hearts. Seven percent had advanced heart block. Is that rare? In serious cases, a temporary pacemaker is sometimes installed for Lyme carditis and occasionally a permanent one if doctors are unaware that Lyme disease is at the root of the abnormal heart function.

Four months after Joseph's death, the CDC published a report of three Lyme carditis deaths in the Northeast within ten months; it did not include Joseph's. Up until then, just four Lyme carditis deaths had been reported worldwide in all of the medical literature. In one of the new cases, a resident of Massachusetts who had complained of feeling unwell for two weeks was found slumped over the wheel of a car. "Abundant spirochetes" were found in the patient's tissues, the report stated. In the second, a Connecticut resident went to an emergency room with shortness of breath and anxiety and went home with an antianxiety medication. He died the next day. "The patient lived on a heavily wooded lot and had frequent tick exposure," the report said.

As a reporter for the Poughkeepsie Journal when the CDC report came out, I ferreted out the identity of the third carditis case, a thirty-eight-year-old father of three and correction officer in training who, like Joseph, lived in Dutchess County, New York. It meant that within thirty-five days, two people had died of Lyme carditis about fifteen miles apart. "He was home in Beacon with his three daughters, six, eight, and seventeen, hanging laundry in the yard on a warm July day," I began the article. "Then, in an instant, he was holding his chest, staggering into the house, pushing furniture aside in a way that made the girls think daddy was clowning around, the way he sometimes did." His wife, who had asked that his identity be withheld, told me the signs were there. "Had he gone to the doctor, he might still be here today," she said. Perhaps. One of the three people in the CDC report had sought medical care, but as in Joseph's case, the signs were missed. I suspect such oversights occur in part because of how Lyme disease is viewed: The tests work. It is overdiagnosed. It is easy to cure.

Of note, donated corneas from two of the Lyme carditis fatalities were transplanted to three patients. One recipient died before being notified that Lyme disease had been found in the donor's heart. The recipient's death was attributed to unspecified "unrelated causes," although no tissues were available for examination. The two other inadvertent recipients were prescribed antibiotics.

A play was written about Joseph Elone, called *The Little Things*, attesting to the power of a tiny tick to wreck lives and upend families. Emmauel Elone was interviewed by my newspaper days after losing the brother who had completed his family. He was eighteen, a boy three weeks from going away to college who was experiencing the toughest chapter in any life. "You should never take life for granted," he said, straining for words. "You should always make sure that you love one another every day. You never know when they're going to go or how they're going to go." He then folded into his father's arms. He and Joseph, he wrote in a blog at Georgetown two years later, were "the closest friends that either of us ever had."

A team at Westchester Medical Center, led by pathologist Esther C. Yoon, wrote a report on Joseph's case, which did not lay blame. "Here, we report a rare example of fatal Lyme carditis in an unsuspected patient," it stated. The case study, published in *Cardiovascular Pathology* in 2015, included this acknowledgment at the end: "We remember Joseph as a kind, gentle, and caring person. Based on his interest in science and his compassionate nature, we believe he would approve of this manuscript to understand the disease that took him away from us."

CHAPTER 10:

Lyme Takes Flight

In 2008, veterinarians in Canada were asked to pitch in on a project with pressing implications for human health. The question: had an anticipated wave of Lyme disease arrived, and where was it emerging? In the United States, where disease-ridden ticks had already spread widely in the Northeast and Midwest, dogs had long served as loyal if hapless sentinels of *Borrelia burgdorferi* infection. Twenty years earlier, Tufts University researchers had found they could use cases of Lyme disease in dogs to predict risk factors for the disease, human and otherwise. Dogs that lived at lower altitudes, namely near the coast, were five times more likely to be infected than others. Sporting dogs, those that romped through fields, were four times as likely. Moreover, and this is where Canada took a page from American Lyme history, the Tufts researchers found they could accurately predict the incidence of Lyme disease in people by looking at rates in dogs.

After collating reports from 238 veterinary practices involving more than 80,000 dogs, the Canadian study indeed found Lyme disease moving steadily, but ominously, north of the US border, at least in dogs. The risk was "low but widespread," the study found, but with distinct areas of higher prevalence. It was these areas, and what the researchers did not report in their data, that intrigued John D. Scott.

In 1990, Scott was working as an agrologist—an occupation involving crop production—when he was diagnosed with Lyme disease in a place where the bug was not supposed to be: his hometown of Fergus, Ontario, a small city on the Grand River in a triangle of Canada bordered by three Great Lakes. That tick bite ended Scott's career, changed his life, and set him on a scientific journey unique in the vast landscape of tick-borne disease research in North America. Over the next quarter-century, he became a leading published scientist on the movement of Lyme disease, birds, and ticks into and within Canada, work for which, he will tell you, he earned "not a cent." His scientific papers list his affiliation as "Research Division, Lyme Ontario," which is an advocacy group for Lyme disease patients. He does not, as many other Lyme scientists do, have the title "doctor" in front of his name or a high-class university position after it. No matter. "Name me chapter and verse where a Ph.D. is required to produce valid science," he said when I asked about his educational pedigree.

When Scott studied that report of 80,000 dogs and the tick-borne diseases they harbored, published in 2011, he noticed something that the study authors had not. The highest rates of infected dogs, he saw, were not along coastlines or near cut up bits of forest that are known to be hotbeds of Lyme disease. Rather, the line of highest infection closely followed invisible aerial highways used by songbirds—the common yellowthroat, golden-crowned sparrow, Swainson's thrush—on their annual north-south migration. As Scott had long believed, birds were dispersing ticks as they always had, but with a new and insidious kick; one called *Borrelia burgdorferi*, the Lyme disease pathogen. As he interpreted the canine data, he saw that the country's migratory flyways were veritable roadmaps for a growing epidemic. The highest prevalence of infected dogs aligned neatly with three of Canada's four migratory bird highways: the Atlantic flyway, running through the Maritime provinces, southwestern Quebec, and southern Ontario; the Mississippi flyway, which passes through northwestern Ontario and southern Manitoba; and the Pacific flyway, which goes north into southwestern British Columbia.

For Scott, a man with a broad, toothy smile, white cropped hair, and firmly held opinions on PhDs and otherwise, it made sense, given all he learned from the birds he had caught, examined, tested, and banded over two decades. Take the Blackpoll warbler he had captured, which flies up to eighty-eight hours straight from Canada to its wintering grounds in Puerto Rico and northern South America, returning the following spring. He had found one with an *Ixodes scapularis* tick firmly attached, one of many firsts he has reported in the scientific literature: First Lyme-infected tick on an indigo bunting. First finding of three tick species sharing rides, or "cofeeding," on the same birds. First spotting of a tick that had never been reported in Canada, *Ixodes affinis,* and another one, *Ixodes auritulus,* never before seen in the Yukon.

In a three-year study, he and two colleagues pulled 481 ticks from forty-two species of migrating birds, from Oregon juncos, spotted towhees, swamp sparrows, and American robins. That the birds were carrying fifteen different species of ticks was one thing. Quite another was what these ticks brought along. Nearly 30 percent of 176 *Ixodes* ticks were infected with the Lyme pathogen. As concerning was this: half of the larval ticks—namely, tick babies that usually hatch clean and pathogen-free—were now infected after taking their first meal. That could mean only one thing: The "larvae almost certainly acquired *borreliae* directly" from the migrating birds, which themselves were "competent reservoirs" of infection. Not only could the birds import ticks into Canada. They could also infect them with the pathogen. "Our results suggest that songbirds infested with *B. burgdorferi*-infected ticks have the potential to start new tick populations endemic for Lyme disease," Scott's paper declared.

Other scientists, like Scott, were taking notice, in particular that birds weren't only carrying the Lyme pathogen. On an island in the Baltic Sea off Germany, popular for migrating birds, researchers from Friedrich Schiller University collected nearly 200 ticks from 99 ground-feeding birds. Seven percent of the ticks were infected with *Rickettsia*, a spotted fever pathogen; 5 percent carried *Babesia*, and 3 percent were infected with *Anaplasma*, they reported in 2007. All of which begs the question: In the global spread of tick-borne pathogens, are birds leading the way?

Ever since the 1980s, when HIV spread around the globe in less than a decade, authorities have worried that hitchhiking germs from far-flung places were merely a plane ticket away. Indeed, the Zika virus, with its potential to cause devastating birth defects, hopped oceans in the mid-2010s in the blood of human beings; those people then infected biting mosquitos in their homelands that went on to spread the virus to other people they bit. Such is the deviously ingenious way of disease transmission. But tick-borne pathogens have their own clever ways of disseminating, geographically and otherwise, that is beyond the reach of any public health travel advisory or warning to wear DEET.

Every spring, about 3 billion passerine birds, including but not limited to songbirds, bring some 50 million to 175 million *I. scapularis* ticks—the ones that impart Lyme disease—into Canada, a government study estimated in 2008. Some birds arrived in Nova Scotia so infested with ticks that researchers posited they had to have stopped along the Atlantic flyway in the northeastern United States, where the ticks have been rampant for decades. Some of the imported ticks came from as far south as Brazil and dropped their cargo as far north as the Yukon. In 2008, the Public Health Agency of Canada mapped the future expansion of ticks and, moreover, of Lyme disease throughout the country. In the previous decade, government and university researchers had watched known populations of *Ixodes* ticks sprout from a single location in the far south of Ontario to twelve more locations—along Lake Erie; on the fringes of Thousand Islands national park; in Nova Scotia; and in southeastern Manitoba. They were bracing for more. Among the data fed into a computer simulation, along with projections of warmer weather, the tally of forested land, and the range of known tick populations, was something called "an index of tick immigration."

Plainly put, a mass migration would deliver more ticks and likely more disease to Canada in coming years, in the form of beautiful waves of song sparrows and wrens, red-winged blackbirds, and warblers of many kinds. And a warmer climate would help these ticks survive in many new places. In a description that sounds something like a page out of a Superman comic, an article in the *International Journal of Health Geographics* stated, "These migratory birds are capable of surmounting geographic features (lakes, sea, mountains and areas of intensive agriculture) that are obstacles to dispersal by terrestrial hosts." Perhaps they aren't scaling buildings in a single bound, but these birds are traversing continents by the billion.

Hopping Mainlands, Crossing Seas

The dispersal of ticks by birds isn't limited, of course, to the North American migratory flyways. In Germany, thrushes, blackbirds, robins, and blackcaps fly the skies from as far away as Sub-Saharan Africa or as near as southern Europe. They sometimes arrive with dozens of ticks tacked around their eyes, ears, and beaks, mostly of the *Ixodes ricinus* or castor bean variety. In a German study of 3,000 ticks harvested from nine hundred birds from 2008 to 2010, researchers computed such things as the share of birds with ticks (3 percent), the average number of ticks per infested bird (3.8); the proportion of ticks infected with the Lyme disease pathogen (about 6 to 9 percent). It is tedious messy work—removing, counting, identifying, and testing arthropods that are sometimes no bigger than a speck of dirt. But it allows scientists to make important deductions. Short distance migrants, for one, were twice as likely to carry ticks, because the arachnids hadn't yet finished feeding and fallen off. Like passengers on modern jets, they fed in the air on a Mediterranean to Germany flight, importing pathogens, the scientists wrote, to "new potential foci." When Gunnar Hasle was working on his PhD thesis at the University of Oslo in the early 2010s, he was impressed by the capacity of birds to cross "mountains, glaciers, deserts, and oceans." It had never been shown that birds had actually seeded a new tick species or pathogen, he observed in 2013, but a case was surely building: "Evidence strongly suggests that this could occur."

At the southern doorstep to the United States along the Gulf of Mexico, similarly, scientists checked 3,000 birds at the border, reporting in 2015 that 3.6 percent were infested with ticks. That translated to 4 to 39 million neotropical ticks carried into the United States annually. Given that almost a third were infected with *Rikettsia*, a bacterium that causes spotted fever, this tick migration, the researchers wrote, poses "uncertain consequences for human and animal health."

Migrating birds have long carried ticks—into Great Britain, where 8 percent arrive infested; onto the Isle of Capri, with one in twenty laden with ticks; and into Finland, where 25 million ticks are imported every year on the backs of 600 million migrating birds.

What scientists believe is different in the twenty-first century is the plethora of spirochetes and protozoans and viruses that they may carry with them. Moreover, these eight-legged menaces, and their unsavory cargo, may be able to survive and thrive in places they never could. "Considering an influx of 30–80 million passerines crossing the sea every spring," wrote Gunnnar Hasle of Norway, "the limiting factor will not be the ticks' dispersal ability, but the suitability of the area the ticks are released for the survival and reproduction of the ticks." Chiefly, will it be warm enough? In many places, the answer is yes.

In the scheme of things, ticks are among the species that will benefit, and greatly, as climate changes. They are moving and populating, increasing their range and numbers in the United States, Canada, Western Europe, the Baltic States, Russia, and China. Mice spread them as they move north. Deer sustain and carry them. Birds import them. In each of those places, Lyme disease is proliferating. But what about South America? What about the place from which birds carry exotic species of ticks that are today being found on the boreal forest floor in the Yukon? Does Lyme disease live there, in the bellies of ticks, in mammals, in nature? The evidence, as scientists are fond of saying, is emerging.

It was not until 2013 that *Ixodes* ticks infected with a species in the *Borrelia burgdorferi* family were reported in South America, first in ticks harvested from cattle, deer, and vegetation in Uruguay. Then in 2014, two more South American countries joined the list. Scientists in Chile tested *Ixodes stilesi* ticks collected from the understory and from long-tailed rice rats and discovered a new branch in the *Borrelia* family tree, named *Borellia chilensis*, for its country of origin. Two months later came a report from Argentina, where the same bug was found in *Ixodes pararicinus* ticks. At that point, at least, researchers from the three countries reported optimistically, "there are no records of *Ixodes* ticks biting humans in the southern cone of South America." We shall see.

Farther north, in the broad expanse of forest, coffee, cocoa, and coast that is Brazil, the first report had already turned up of what came to be called Lyme-like illness or Baggio-Yoshinari Syndrome, named for the scientists who identified it. In 1992, two tick-bitten brothers living in the State of São Paulo in the country's southeast became the first victims of a disease that researchers, aware of an unfolding outbreak in the US Northeast, had been searching for since 1989. While the disease came with rashes and arthritis-like symptoms and bore some similarity to Lyme disease, its pathogen stubbornly refused to reveal itself. In 2007, a scientific article bore the intriguing title: "Description of Lyme disease-like syndrome in Brazil. Is it a new tick-borne disease or Lyme disease variation?" In 2010, a rather frustrated research group from São

Paulo recounted, "In laboratorial terms, bacteria from the *B. burgdorferi sensu lato* complex"—the broad family of Lyme bugs—"were not isolated in biological fluids and tissues whatsoever." By 2016, researchers knew only that sick patients harbored antibodies to *Borrelia burgdorferi s. l.*, but that the pathogen could not be cultured and appeared in patient blood without the distinctive *Borrelia* flagellum. Then in 2017, there was a breakthrough. Using DNA testing, researchers found the genetic footprint of *B. burgdorferi* in four women from the southwestern state of Mato Grosso do Sul. One was ill with early Lyme-like illness, the others with advanced arthritic, cognitive, and psychological manifestations.

What was ultimately named Brazilian borreliosis was different from the North American version of Lyme disease, which itself differed from Europe's Lyme borreliosis—a function of varying species and strains. Brazil's disease is different in other ways too. The prime "reservoir" from which ticks are infected is not a tiny mouse or assorted other mammals but a 100- to 150-pound rodent called a capybara that likes to swim, has coarse reddish fur and a broad blunt snout, and was spotted on the fifth hole of the Olympic Golf Course in Rio de Janeiro in 2016. Further, it remains unclear which tick family is the prime vector—*Ixodes, Amblyomma,* or *Rhipicephalus.* In that respect, Brazilians share a problem with residents of the American South, whose Lyme-like illness is delivered by *Amblyomma,* or lone star ticks, and is called Southern tick-associated rash illness. The pathogen that causes it, its relationship to Lyme disease, and the possible role of *Ixodes* ticks are still unknown. Clearly, we don't have this whole thing figured out.

Beyond the role of birds and ticks and assorted hairy mammals, South America's experience with tick-borne disease is instructive in another way. As in North America, the malady "was reported to cause neurological, cardiac, ophthalmic, muscle, and joint alterations in humans," and it responded well, early on, to antibiotics. But relapses there were common and harsh, described in the *Brazilian Journal of Microbiology* in 2017 as "symptom recurrence, severe reactive manifestations such as autoimmunity, and"—in a statement reminiscent of the early days of Lyme disease in the United States, "the need for prolonged treatment." Will Brazilian doctors someday be censured for treating patients with additional courses of antibiotics, which they presumably are finding helpful, just as American and European doctors?

In São Paulo, where two young boys became the first Lyme-like cases in 1992, a plague hit in 1918. In the course of ten weeks, 100,000 of 500,000 residents were stricken with influenza. More than 5,000 died. Lyme and tick-borne disease surely do not work that way. We can all be grateful for that. But their slow and even spread, their complex ecology, and their broad array of symptoms has made them a more difficult target. Brazil, which is almost the size of the United States, is yet another new frontier.

Shorter, Earlier Journeys

Charles Francis has charted the ebb and flow of birds in Canada for fifteen years as manager of bird population monitoring for the Canadian Wildlife Service. He has used marine radar to gauge the densities and flight patterns of migrating birds over the Great Lakes. He has studied threats to birdlife as large as habitat loss and as specific as wind turbines. It was Francis who, in chapter 1, was on a team that pulled the tail feathers from migrating gray-cheeked thrushes to learn that they were en route to the taiga and boreal forest of the southern Canadian Arctic; the molecular composition of those feathers signaled to scientists that the birds had been born there. Francis and his colleagues had found, moreover, that the thrushes sometimes carried ticks to these far-flung places, supporting the hypothesis that *Ixodes* ticks may just be spreading north on the backs of birds.

That climate is changing fundamentally—and that it is changing nature—is a guiding principal of Charles Francis' work as Canada's bird manager-in-chief. And he worries and wonders. What of those birds that forget or fail to migrate—the ones still around at Christmastime when they would normally be in Mexico? Will an altered migratory clock leave them enough to eat in their springtime homes? Will they survive?

In 2005, Pete Marra, head of the Smithsonian Migratory Bird Center in Washington, DC, devised a way to measure the effect of weather on birds migrating along the Atlantic flyway. Made possible by the dogged work of avid American and Canadian bird monitors, Marra, Francis, and other researchers scoured the diligently kept records from 1961 to 2000 of birds captured and tagged as they migrated from a birding station in Louisiana in the southern United States to two stations about 1,500 miles north in Pennsylvania and far southern Ontario. Other researchers had already tied climate warming to all manner of changes for birds: earlier breeding and extensions in range, to name two. In one study, researchers had found a mismatch between the hatching of baby birds at Vic-le-Fesq, thirty miles north of the Mediterranean in France, and the availability of food. As a paper published in the journal *Science* put it, adult birds were pushed "beyond their apparent sustainable limit," trying to feed babies that arrived before caterpillars were ready.

Marra et al. added to that body of knowledge. For every 1 degree Celsius increase in spring temperature, they observed, birds arrived a day earlier on their annual migration north. Further, these avian migrants had not left their tropical homes earlier but rather made the made the trek through eastern North America faster in warmer years. The study made the case: shifts in temperature meant changes in migration, suggesting what future wholesale changes in climate might do and, in other places, already had done. In the United Kingdom, three studies showed migrating birds arriving up to two weeks earlier in a twenty- to thirty-year period. Two dozen studies in Europe and Asia demonstrated that migrating birds arrived 3.7 days earlier on average with every passing decade. In the Southern Hemisphere, meantime, the inverse has unfolded. Hampered by a shortage of sea ice in East Antarctica, penguins and petrels have delayed breeding cycles that had been set in ice, if not stone, for generations: they arrived at breeding grounds nine days later, on average, and laid their eggs two days later than in the early 1950s.

But Is This New?

In Canada, John D. Scott and his colleagues were busy from the mid-1990s through the 2010s identifying ticks pulled from migrating birds. In 2005, he reported the first instance of an *Ixodes auritulus*, or avian coastal tick, infected with the Lyme disease pathogen on Vancouver Island, British Columbia; it had hitched a ride on an American robin, originating somewhere in Central America. At Watson Lake, which is above the 60th parallel north in the chilly Yukon Territory, he found three new species on long-distance flyers: *Ixodes brunneus, Ixodes muris*, and in 2012, another *Ixodes auritulus* that had never been seen so far north. He also identified the first lone star stick, *Amblyomma americanum*, from a Yukon migrant in 2010, and in 2015, two other types of *Amblyomma* tick on migrating veeries. Scott's report included this observation, fascinating if one appreciates the power of a pair of wings on a two-ounce frame: "Veeries could theoretically transport *A. dissimile* [an *Amblyomma* tick] from as far south as southeastern Brazil, a distance of over 7,500 km." That is roughly 4,600 miles.

Scientists still do not know precisely what all this adds up to. The lone star tick, among the new species found by Scott, cannot overwinter in Canada, he said; it's just too cold—for now—for their larvae to survive. An American study, recall from chapter 8, predicted significant northward movement of the lone star tick, from the American Southeast to huge stretches of the Great Plains and Midwest. Will it move farther north still?

While large questions remain, scientists are confident of this: *Ixodes scapularis* ticks, laden with Lyme, will pour into Canada on birds from the United States, where Lyme disease is well established, a phenomenon playing out south-to-north in other countries around the world. Those birds will fly the Atlantic flyway from the US Northeast to the Atlantic provinces. They will take the Mississippi flyway from the US Midwest to western Ontario and points west. To a lesser extent, they will cross the border from western areas of Pennsylvania, New Jersey, and New York and into eastern Ontario and Quebec. In 2013, government researchers released a study of what it called a current and anticipated tick "invasion," a word it used thirty times. In southern Quebec, the stage was set in 2004 by a cluster of ticks, new to the region and the country, which by 2009 had reached a critical mass, what researchers called, "the establishment of efficient *B. burgdorferi* transmission cycle."

That study gave the country five years from the seeding of new tick populations to the point at which "significant Lyme disease risk emerges." "Different parts of Canada will receive ticks carried by migratory birds from different parts of the USA," the study predicted. Other countries, too, may see ticks take root that formerly could not survive. For now, for example, the Ornate cow tick, which carries pathogens that cause spotted fever and tick-borne encephalitis, cannot overwinter in Norway. It is nonetheless a champion survivor, able to live for months under water and "to overcome years of unfavorable conditions," according to a 2015 report in the journal *Parasites & Vectors*. The report designated four of five Scandinavian countries (Denmark excluded) as places where the tick was "anticipated-absent." In other words, this "vector on the rise," as the tick was called, is not there now. But it's likely coming.

Scott, the scientist who has done more than most anyone to document sightings of ticks from faraway places, thinks these arthropods— whether exotic hitchhikers from South America or run-of-the-mill ticks from the neighbor down south—have likely been carried into Canada for thousands of years. He disputes assertions that a warming climate is creating new places for more ticks, including the indigenous *Ixodes scapularis* tick, the primary carrier in North America of Lyme disease. In Canada in 1911, two researchers named Nuttall and Warburton reported the first *Ixodes scapularis* tick on a person in Bracebridge, Ontario, 200 miles north of where the ticks were said to have emerged nearly a century later. The ticks have always been there, he maintains, and his argument makes a certain amount of sense: ticks adapt. They have a kind of antifreeze compound in their bodies that allows them to withstand intense cold. The climate has changed in Canada, sure, but what's a degree Celsius of warming to a tick?

"Climate change doesn't have one iota of effect on it. To me it's just a bunch of baloney," Scott told me more than once. "Birds are dropping these ticks all over the place. It's been going on here in Canada for the last 10,000 years since the last Ice Age." When Scott contracted Lyme disease thirty years ago, his hometown of Fergus, Ontario, where he believes he was infected, supposedly had no infected ticks. Government officials insisted into the late 1990s that the only place to get Lyme disease was in Long Point, Ontario. Climate change and ticks? A fad to get research grants, by Scott's way of thinking.

I have spoken with other scientists like him who believe we may simply be finding ticks because we have a reason to look. Indeed, historical data is scant, as I covered earlier. But what little there is points to more ticks in more places. The better question may be, perhaps, are we finding more ticks with more disease? That is a trend Scott does embrace. In 2006, 13 percent of adult *Ixodes scapularis* ticks across Ontario were infected with the Lyme disease spirochete. A decade later, Scott reported an infection rate of 73 percent in the adult ticks on Corkscrew Island in sprawling Lake of the Woods in southern Ontario. He attributes some of the huge increase in infected ticks to the technology used to detect the pathogen. But still. "What was once considered by some researchers as a hostile environment for *I. scapularis* has turned out to be one of the most hyperendemic areas for Lyme disease in Canada," he and his colleagues wrote in that report in 2016.

In the 2000s, researchers ventured to Gull Island, off the coast of Newfoundland, where they checked seabirds for ticks, finding, not surprisingly, that ten of sixty-one *Ixodes* ticks were infected with the pathogen that causes Lyme disease. What was surprising was that the Lyme pathogen found was not the common North American species. While this species, *Borellia garinii*, had been known to exist in Europe, Asia, even Alaska, it had never been seen on the continent's east side. I wrote earlier about a Canadian woman, Sue Faber, and her young daughter, who were both infected with European strains of Lyme disease that standard tests did not catch. They are harbingers perhaps. In 2014, a report from Canada's National Collaborating Centre for Infectious Disease said attention should be paid to the Gull Island findings: "This could introduce *B. garinii* into Canadian mammalian populations, and cause Lyme borreliosis in North America with significantly divergent symptoms."

Here, then, is a new wrinkle in the avian importation picture. While the general trend of ticks and disease is northward, this pathogen has the potential to be carried east and south too, a newcomer to Canada and, potentially, the United States, courtesy of the wings of a few million migrating birds.

CHAPTER 11:

A Lyme-Free World

From late 2015 through 2016, a mosquito-borne virus named for the Zika Forest in Uganda was linked to 2,336 cases of Brazilian babies born with horrifically undersized heads, a condition called microcephaly. As reports accumulated, the threat of the Zika virus nearly shut down the 2016 Summer Olympics in Rio de Janeiro. Then the virus showed up on the shores of Florida. The US government response to the Zika threat was swift. By September of that year, Congress approved, and the president signed into law, a $1.1 billion commitment to fight the mosquito-borne threat. Just three months later, the US Centers for Disease Control and Prevention awarded $184 million in grants for Zika monitoring, prevention, and research. That same year, in sorry contrast, the CDC awarded $2.7 million in grants to control Lyme disease, the second-leading infectious disease in America.

Zach Adelman is a forty-something mosquito researcher at Texas A&M University who had been very busy in the months after Zika exploded into the New World's consciousness. Adelman's lab was flush with funding from the US government, and it was developing a method, not of any small consequence, to defuse the power of disease-carrying

Aedes aegypti mosquitoes. The breakthrough would not only curb the threat of Zika but potentially of other devastating mosquito-borne diseases, like dengue fever and malaria. When I spoke to him in the spring of 2017, Adelman and his colleagues were on their way to genetically engineering mosquitoes that would, if scientific and ethical hurdles were cleared, produce male-only offspring. Within a few years perhaps, a nuisance bug that had caused millions of deaths over thousands of years could, when and where necessary, be vanquished. Zika, discovered in a rhesus monkey in Africa in 1947, was merely the latest addition to *A. aegypti's* extensive and lethal cargo. Its power to inflict misery was predicted to grow as climate change ushered in new pathogens and expanded the insect's geography. If anything, the grants aimed at the mosquitoes were inadequate and overdue, given the global implications of this ubiquitous bug.

Mosquitoes have far shorter lifespans than disease-toting ticks: about two weeks versus two years. Adelman's technology could quickly dispatch mosquitoes, even if it involved several generations to take effect. Ticks, however, live mainly in a state of Rip van Winkle-like repose, waking only for three blood meals, one in each life stage, over the course of perhaps eight seasons. Applied to ticks, Adelman's idea would take literally years to become established. But making male-only ticks could theoretically be done, Adelman told me. Except for one thing. Money. Unlike Zika, Lyme disease doesn't have it, doesn't get it, isn't showered with it.

Yet Lyme disease produces pain and sometimes profound disability, as with Zika, and likely on a bigger scale. As of April of 2017, Zika's American count since the outbreak began in 2015 stood at 224 infections that were contracted in the United States and about 5,000 from travel to other countries. Ninety-one infants were born in the United States with Zika-related birth defects through July of 2017, and eight more died in utero. By contrast, sixty-three American deaths were formally attributed to Lyme disease in 2014. About 300,000 to 400,000 people are infected in the United States annually, with 10 to 20 percent of casualties developing long-lasting, recurrent complications. Both illnesses can be serious but can also be fleeting and inconsequential, especially so for early, adequately treated Lyme disease.

This isn't about which disease tops the other as a public health calamity. It is about whether responses to each are proportional. Zika prompted panic and got attention. Lyme disease, by contrast, has long been so much background noise, acknowledged but minimized by public health officials and the researchers they relied upon. In 1991, the *Annals of Internal Medicine* published an article mocking Lyme disease patients entitled, "From the Centers for Fatigue Control, Weekly Report—Lime Disease, United States." "Lime disease" occurred more frequently among middle- and upper-class people and women, the article snidely reported, with studies showing a strong link to reading stories about Lyme disease.

In 1998, and in the same vein, a proposed Lyme vaccine was famously referred to as "as yuppie a vaccine as I've ever heard of" by a Duke University pediatrician. The comment was made at a meeting of a committee that advised the CDC on vaccines. The committee voted against recommending the vaccine in high-risk areas, opting instead to give the highly unusual advice that it "should be considered." Sixteen years later, a committee member reflected on that regrettable moment. "By calling it a 'yuppie vaccine,' by damning it with the faint praise of a 'should be considered' recommendation, we killed it," Paul Offit, chief of infectious diseases at the Children's Hospital of Philadelphia in Pennsylvania, told the journal *Nature Medicine*. "Some on the committee seemed to view the vaccine as a luxury for the anxious affluent rather than a public health necessity," the article recounted. Indeed, this is part of a pattern that has persisted for many years. A researcher I quoted earlier said he was denied funding with the comment by a reviewer that this was a "middle-class disease."

The $2.7 million that the CDC gave out in Lyme grants in 2016 was less than 1 percent of the agency's research awards that year, a figure even I was shocked to learn. When I questioned this in the spring of 2017, the CDC referred me to its "Ongoing Research" website page. Of the six projects listed, three had already been wrapped up, including two several years earlier. It was hardly a robust list of active research, which, on the day I raised the question, was quickly relabeled, "Ongoing and Published Research." Zach Adelman, the mosquito researcher on the brink of tackling a global public health scourge has a theory on the paltry funding for Lyme disease. "The symptoms are not as heart-wrenching," he explained. "They don't photo as well. You don't have hemorrhagic fever. You're not dripping blood out of your eyes. You don't have deformed babies. But," he said of Lyme disease, "it's the same thing." In other words, the pain of Lyme disease can be as intense, the damage as significant.

Ironically, in the early years of Lyme disease research, there were reports in the scientific literature similar to those of Zika, of babies and fetuses infected with, and potentially damaged by, Lyme spirochetes passed by their mothers. Among sixty-six "adverse outcomes" involving gestational Lyme disease, assembled for a 2001 textbook on infectious diseases in children, twenty-six involved death in utero or shortly after birth. In these studies, spirochetes were found in the hearts, brains, and kidneys of such babies, establishing, as a 2001 literature review in the journal *Teratology* put it, "*B. burgdorferi* can cross the placenta." The Lyme organism was found in a stillborn baby with a heart defect and in six second-trimester fetuses, three with cardiac defects. But because researchers had been unable to document inflammation—namely illness—in fetuses and babies, and the defects were highly varied, gestational infection has largely been dismissed. Researchers, unwisely, moved on. Many ill mothers have told me they believe they gave Lyme disease to their children.

In August of 2016, NPR reported, "Fourteen people likely caught Zika in a neighborhood north of downtown Miami....That means mosquitoes in that area have picked up the virus and are spreading it." The Zika threat nonetheless quickly waned, likely from "herd immunity," in which Zika-infected people develop antibodies against future infection and the virus dies off. No such phenomenon exists for Lyme disease, which also lacks images of insects that fly and infants destined for lives of dependency. The same week NPR told of those 14 Zika cases, 331 Lyme disease infections were reported in the United States, perhaps a fifth of which would lead to lingering cognitive, neurological, or joint problems.

In the late 1970s, Stephen Wikel, a newly minted PhD in veterinary microbiology from the University of Saskatchewan, made a series of observations about ticks that, had the stars aligned, might have stopped the modern-day epidemic of tick-borne disease in its tracks. He was the Zach Adelman of his generation, with an elegant and promising idea but without, as it turned out, the money.

Four decades ago, Wikel found that the more times a guinea pig or rabbit was bitten by a tick, the more sensitive the animal became to the bite. Moreover, ticks fared far worse when they fed on animals that had been bitten before. They didn't grow as fat, weren't as fertile, laid more impaired eggs, didn't molt as efficiently, and died sooner. What was at the root of this acquired resistance, he thought? Somehow, these animals had begun to develop ways to fight off some of the magical properties of tick saliva, the ones that subverted normal host immunity, numbed skin, and prevented blood clotting and healing at the bite site, and allowed ticks to feed to their hearts' content. If that process could be identified, Wikel thought, if it could be artificially simulated in would-be tick hosts—in people—science would have found a means to prevent infection. This would not merely be a way to prevent one tick infection like Lyme disease. This would ward off ticks altogether.

"Block the key molecules that allow for the creation of this privileged site for the pathogen to be dumped in." That was how Wikel described the idea for me for what is called an antitick vaccine. Having blocked those molecules, and disarmed the tick's all-powerful saliva, the human immune system could do its job against an invader on newly hostile territory. Ticks would not be able to plant a toehold. They would fall off, and perhaps even die. Wikel chiseled away at the science behind this idea for several decades, writing many papers on the "cutaneous interface" between tick and host. He contributed greatly to the literature but retired without a vaccine. Government interest, once enthusiastic, waned. Funding dried up. Ever committed, however, Wikel still has hope.

Short of restoring balance to a wounded planet, there are two options to curb Lyme and tick-borne disease. First, get rid of or, more practically, sharply reduce ticks. Second, stop them from infecting people. We are a long way from reaching either of those goals. It is a problem of will, not ability. Science has tackled bigger problems and in less time.

The Lesson of AIDS

Mayla Hsu became a scientist in the 1990s, in an era when the line charting infections with HIV and deaths from AIDS had been on a steep upward climb for more than a decade. A student at McGill University in Montreal, she was compelled by the challenge of helping unravel the cause and cure of a devastating health scourge. Those were heady days. There was money for fellowships, conferences, and research. The National Institutes of Health provided reagents to speed experiments and assure consistent results. Research institutes supported a trove of mentors and peers with whom to work and grow. Grant by grant, lab by lab, PhD by PhD, an infrastructure was built to find an answer. Science did.

When I spoke to her in the spring of 2017, Hsu had left a productive career in HIV research where she had published successfully on, among other things, the ways the organism was able to evade drug therapy.

Now in her midforties, she had turned her attention to Lyme disease as research director of a nonprofit organization, Global Lyme Alliance, one of several donation-driven organizations that were funding Lyme disease science. In contrast to HIV, Hsu saw in Lyme disease a research milieu starved for government grants, lacking encouragement for new researchers, and painfully short of the resources that had minted a generation of scientists like hers. "There is no career path for these people that want to study Lyme disease," she told me. Significantly, Lyme disease was also of little interest to the pharmaceutical industry, she found, which had seen money to be made in HIV drugs. Hsu called Lyme disease not a parallel to HIV but "an anti-parallel."

Zach Adelman, the mosquito researcher at Texas A&M, was witness to Lyme disease's second-class status when he sat on a government panel to review applications for funding. There, scientists from many disciplines were brought together to decide the future of government research. But however big the Lyme problem was, he found the community of tick researchers represented there to be small and subordinate, mostly because their science had not been funded in the past. Hence, grant proposals for tick-disease research were drowned out by better-funded interests, in something of a cycle that perpetuates itself, interrupted only by a breakthrough disease, like Zika, that changes priorities. It doesn't help that leaders in treatment policy have portrayed Lyme disease as an easy disease to diagnose and treat, while minimizing its devastating impacts on many.

In 1982, Congress allocated the first $12 million for AIDS research and treatment. That was a year after the Lyme disease organism, *Borrelia burgdorferi*, was identified in a US government lab and the year before the AIDS virus was identified. But that is where the two epidemics diverge. By 1997, AIDS research money to the National Institutes of Health amounted to $1.5 billion. That year, the AIDS death toll, which had reached its apex in the United States of 41,699 in 1996, began a precipitous decline. A standard of care had been discovered that turned HIV from a fatal to chronic infection. The reason was money, lots of money, poured into an epidemic that was crying out for it. The lack of such funding for Lyme disease is why Stephen Wikel's idea for a vaccine against tick saliva remains a great idea waiting fulfillment. The disease lacks the infrastructure that turned Mayla Hsu—and many more like her—into a microbiologist and immunologist in the era of AIDS.

In December of 1998, the first vaccine against Lyme disease, called LYMErix, was licensed by the US Food and Drug Administration and marketed by the pharmaceutical company SmithKline Beecham. The vaccine's success was limited, and its commercial life short. For one, even if people got all three doses—a cumbersome hurdle to immunity—just 78 percent would be protected. That figure dropped to 50 percent if two doses were received. For another, and most crucially, the vaccine was said to have elicited Lyme-like arthritic symptoms, which was the apparent death knell for LYMErix. A study of adverse events, published in the journal *Vaccine*, found no "unexpected or unusual patterns" after the first 1.4 million doses had been administered. Nonetheless, it dutifully reported, there had been 101 reports of various kinds of arthritis and sixty-six adverse events that involved "life-threatening illness, hospitalization, prolongation of hospitalization, persistent or significant disability/incapacity, or death."

SmithKline pulled the vaccine from the market the same month the study came out in early 2002, citing poor sales. I can't say whether the vaccine was unsafe. But a Lyme-only vaccine is inherently flawed because it does not address a key problem. Ticks carry more than Lyme disease. The week I wrote this, a study was published in the journal *mSphere* showing 30 percent of adult ticks from Long Island, New York, infected with *Babesia microti*, which causes babesiosis, on top of 67 percent that carried the Lyme pathogen. Some also harbored *Borrelia miyamotoi*, for which there is no diagnostic test, and deadly Powassan virus, with a 5 to 10 percent fatality rate. And while their pathogen loads are growing, ticks are also spreading widely and in huge numbers. No one should think they are protected with a vaccine that targets *Borrelia burgdorferi* only. It is something, to be sure. It is better than nothing—I would welcome it in fact—because Lyme disease makes other infections, like babesiosis, worse. But we need a better vaccine than one that only tackles Lyme. We need more protection.

In Europe, a concerted effort is being made to do that, to build a vaccine that works against ticks—Stephen Wikel's dream come true—and not only against Lyme borreliosis, as it's called there. Seven institutes in six countries have been funded by the European Union in a project called ANTIDotE, which loosely stands for Anti-tick Vaccines to Prevent Tick-borne Diseases in Europe. In the United States, scientists at Tufts, Yale, and other institutions are also doing basic research that could lead to a vaccine; some of their science is integral to the effort. But there's nothing akin in the United States to the coordinated, European Union-funded approach.

Europe has been in this place before, after all, and it has prevailed. Consider tick-borne encephalitis, or TBE, a virus common in ticks from the fringes of France in the west of Europe to Asia in the east, from Albania in the south to Russia in the north. The disease causes brain swelling and sensory problems in 20 to 30 percent of infected people and can lead to long-term psychiatric problems or even permanent neurologic damage. About a week after onset of neurologic symptoms, the virus kills one or two of a hundred infected people. That kind of injury and death toll may explain why the first vaccine for the disease was unveiled in 1941, just four years after the organism was identified. It suggests why there is concern now in Europe for the growing potential of ticks to inflict harm.

Lyme disease and TBE are similar in many ways, with a key exception. TBE's fatalities occur quickly and can be documented. Lyme disease's true mortality rate is unknown. But there are perhaps a hundredfold the number of Lyme borreliosis cases in Europe today as TBE. In the Netherlands, the country leading the ANTIDotE project, 1.5 million people are bitten by ticks annually, nearly a tenth of the population, resulting in 25,000 Lyme cases in one tiny country. Research there is showing an increase of infected ticks, a longer tick season, and growing areas of tick habitat. Just as in many other parts of the world.

The Wonders of Spit

To a one, tick scientists I spoke with for this book were in awe of, were humbled in fact, by the survival tactics of ticks and, in particular, by what's contained in tick saliva. "The success of their life strategy can be attributed, in part, to saliva," wrote a group of American and Czech researchers in 2016, in an article with the words "Spit-acular Saliva" in the title. Tick spit holds an "armamentarium" of molecules, a Czech scientist named Michalis Kotsyfakis told me, each to carry out specific tasks. The challenge is figuring out which molecules of perhaps thousands serve what purpose—some are anticoagulants, some thwart signals between human immune cells, some stop production of them, others facilitate feeding. Then, the trick is identifying other molecules that may be waiting in the wings, like reinforcements in combat. Knock one out of commission with a vaccine intended to set off an attack of antibodies, and a handful more step up to do the job.

"We need to find the cornerstone," the Greek-born Kotsyfakis told me in a Skype call from Prague. In fact, he added, "We try to find two or three cornerstones to make everything collapse." Kotsyfakis, a bearded, slightly balding man who worries for people who don't understand the threat of ticks, began studying tick saliva as a postdoctoral fellow at the US National Institutes of Health in the early 2000s. He worked for five years under the mentorship of Jose Ribeiro, a government scientist legendary for his dedication to unraveling the tick mystique. Kotsyfakis now heads the Laboratory of Genomics and Proteomics of Disease Vectors at the Academy of Sciences of the Czech Republic, where he dissects ticks at the level of genes and molecules. He and his colleagues, part of the EU's ANTIDotE project, have identified a handful of "pluripotent" proteins that might be used as the basis of an antitick vaccine. "Understanding the molecular mechanisms that govern the life cycle of ticks," they wrote, "is within grasp."

The ultimate goal of the vaccine is for a hungry tick to imbibe on blood laced with inoculation-induced antibodies that would interrupt feeding, cause ejection, or best case, kill the tick. The vaccine might also work in other ways depending on the molecules researchers can identify and target. It might, for one, halt pathogen transmission without stopping feeding, which isn't ideal. In another approach, it might trigger a response to a biting tick—a noticeable itch or irritation—in line with the "acquired resistance" that Stephen Wikel found in his guinea pigs. A would-be human host, normally helpless against something it cannot feel, would then rid itself, like opossums are known to do, of the parasite.

A research colleague of Kotsyfakis', Petr Kopáček, resides in the small town of Budweis, south of Prague, where he routinely picks ticks from the mane of his Australian shepherd, Andy. He likes the "acquired resistance" approach, because, as he is fond of saying, "Humans have one advantage. We have hands." As in, if we feel ticks, we can pick them off. He also believes this route may be the speedier, more promising one. Right now, he feels like he is in something of the slow lane, and it's clear why.

In his laboratory in the Institute of Parasitology, Kopáček works with *Ixodes ricinus* ticks, the European relative of the predominant North American tick, *Ixodes scapularis*. While the magic of modern gene sequencing has opened many doors in tick research, the process is nonetheless protracted. First, Kopáček lets larval, or baby, ticks feed on both infected and uninfected mice. Then the ticks molt into nymphs, from which Kopáček removes the salivary glands. Think surgery on the head of a pin. Then he sequences the RNA in the saliva to see which molecules show up in infected ticks—are "up-regulated"—and which are missing, or switched off, in clean ones. Molecules are thereby identified that may be key to tick survival, the ones that possibly drive successful attachment and feeding.

Once identified, Kopáček uses a process developed by two Nobel Prize-winning biologists called "RNA interference," or RNAi; it permits scientists to switch off selected genes on the molecule to see how that affects the biting-fixing-eating process—to confirm the molecule's role. Then the experiment moves toward its ultimate goal: using the protein molecule—in a recombined, purified form—to make a vaccine that is tried first in animals. Will the animals be protected against infection when tick-bitten, or will they get sick? This is a bit like the search for the Holy Grail, an odyssey involving literally several thousand genetic samples that must be tested. "It is still a lot of work before a candidate molecule may became a vaccine," Kopáček said, who frankly admits he is skeptical of the chance of identifying the right molecules in tick spit. Moreover, he doesn't trust the organism at the center of this. *Borrelia* has learned to survive in cold-blooded ticks and warm-blooded mammals and to pass effortlessly back and forth, in Kopáček's words, "to create an efficient camouflage" everywhere it goes. "The *Borrelia* is very clever," he told me.

Kopáček's colleague, Kotsyfakis, was similarly reserved about the chances of success, not so much because of the science but because of the support. A victory over *Borrelia*, and all the other organisms carried by ticks, will require much more than ANTIDotE promises. The EU's largesse is too small. The US commitment is anemic, and its approach, disorganized. The mission is complex and daunting. "I feel really that this is a neglected disease," he told me. "In the Czech Republic, we have the encephalitis virus,"—Tick-Borne Encephalitis or TBE—"people die almost immediately. Whenever you try to apply for Lyme disease

funding, even in the states, any kind of funding organization, they consider the ticks do nothing because you will not die."

Beyond the obstacles of funding and political support, there is
another challenge to a vaccine that targets ticks themselves. They must
still bite. The Powassan virus, named for the town in Canada where it
claimed its first victim, a child, can be passed in as little as fifteen minutes. Could any antitick vaccine work fast enough to muster antibodies
in the time a Powassan-laced tick did its dirty work? Does that negate
the value of an antitick vaccine? I asked this of Thomas Mather, known
at the University of Rhode Island as the "Tick Guy," who has fought
Lyme disease in the field and the laboratory for decades. "There's always
these little one-offs"—rare though Powassan may be—"that sort of spoil
the concept," he told me. But the idea holds promise. In 2014, Mather's
team reported success in sensitizing what he called "humanized" mice to
tick saliva, thwarting half of Lyme infections. He foresees a time when
that approach might work against other pathogens.

In the meantime, Mather brings a practical bent to the huge problem of ticks, a chip-away-at-the-edges, we're-all-in-this-together, bet-
ter-do-something approach. He believes people should let go of the idea
that science is going to solve the problem any time soon. As part of his
self-empowerment theme, Mather periodically has folks in the community give him their shoes. He sprays them with permethrin, a synthetic
form of an insecticide produced by chrysanthemums, and happily gives
them back. His own study showed that folks in permethrin-treated
socks and sneakers were seventy-four times less likely to be bitten by
ticks. "It's easy, and it helps," he said. Call it one bullet in the war on
Lyme disease. Mather would gladly take an antitick vaccine, however
much it may occasionally miss a quickly transmitted pathogen or two.
"It's not perfect," he said, "but better than where we are right now."

Where are we now? On a mid-April evening in 2017, I ventured
onto the path that rings the field across the lane from my home.

The breeze was gentle, and the air, warm and inviting. I returned from that walk with my husband, three family dogs, and twenty-five hungry blacklegged ticks. I foreswore future walks, the first time I had done so in thirty-five years. Ticks, nonetheless, have ways. Two weeks later, after a morning shower, there was this itch, at the very top of my inner thigh. Something was stuck in the skin in my groin, so small yet protruding like a tiny upright soldier that I knew instantly. I'd been so careful. I had checked myself. I had banned the Shih tzu from the bed to avoid imported arachnids. Two days before, however, I had taken a walk on a rail trail and picked the dog up afterward. That's all I could figure.

I proceeded to pull the thing out, using a fine-tipped tweezer to grab it sideways by its neck. The trick was not to squeeze the contents of the tick's gut, potentially laden with pathogens, out of it and into me. Then I placed it on a bit of tissue, took a cell phone picture and blew it up. There it was: An engorged nymphal *Ixodes scapularis* tick, fat and filled with my blood even if it was a mere speck next to the paper clip I'd included for scale. I thought of the many things I'd read and written on what's in ticks: *Borrelia burgdorferi. Babesia*, akin to malaria, prominent, growing, and seeping into the blood supply. *Anaplasma phagocytophilum, Borrelia miyamotoi, Bartonella henselae.* Powassan virus, found in 5 percent of Wisconsin ticks and in 1 to 5 percent in the Hudson Valley. I looked up the odds. In the Netherlands, 2.6 percent of physician-examined tick bites led to Lyme disease, a Dutch study had reported in 2015. But that only counted people who had manifested the erythema migrans rash; in the United States, some 20 to 30 percent of cases do not. And it did not consider those other pathogens carried by what Dr. Kenneth Liegner, an early Lyme pioneer in New York, aptly described to me as "cesspools of infection."

Between the time of the walk in the field and the tick in the shower, I called a company in the New York region that helps rid properties of ticks, using a promising technology tested in research underwritten by the Centers for Disease Control. First, the company sprays the backyard's brushy edges and leaf litter with an insecticide. This is something I've not done in forty years as a homeowner but am growing ever-more open to. Then it sets up small CDC-tested feeding stations or "bait boxes." Here, mice pass through insecticide-coated brushes before the reward of a tasty meal. The chemical thwarts ticks from feeding off, and becoming infected by, the mice. Yards are made tick-free. The estimate to treat my .69 acre was $840; for my son's one-acre homestead, $1,200.

My family is very important to me. But this, clearly, was not the answer, for me or for society at large. At those prices, a relatively infinitesimal number of ticks will be killed off on isolated properties occupied by well-heeled people. And mice, chipmunks, birds, and so forth will eventually bring the ticks right back in.

At an animal preserve near where I live, helmeted guineafowl roam between pens of goats and sheep, darting, pecking, and, rather unpleasantly, voicing calls described as "buck wheat" for the hens and "chee chee" for the cocks. The birds can also be seen on a few local lawns and at a nearby horse farm, part of what a 2006 paper in *BioScience* called their "cult status" as a natural tick remedy. We need environmentally friendly tick solutions, to be sure. The problem is the birds only eat adult ticks and mainly in lawns, researchers have found. That leaves many smaller nymph ticks, particularly in brushy areas, and they are the ones responsible for most cases of Lyme disease. Research is more promising on natural, plant-based, tick-control substances like nootkatone, an oil from Alaska red cedar wood; rosemary oil, and 2-undecanone, a compound in, among other things, wild tomato plants. The CDC gives these substances tacit endorsement by posting favorable research on its website. But consumers must muddle through complex science to figure out what to choose. They need guidance. They need a clear path. They need protection.

Superbot: Let Ingenuity Loose

In the mid 1990s, Holly Gaff was a mathematician at Old Dominion University, in Norfolk, Virginia, who wanted to quantify the statistical risk of tick-borne disease. The problem was she had little data with which to work. Following the literal path before her, Gaff began collecting ticks, helping catalog about 100,000 ticks from which to learn. To mathematically model risk, you need to know how many ticks are out there and what is in them.

By 2014, Gaff had delved so far into the woods that she had become a biologist who had at least some of her answers on tick-borne risk. That's when she authored a paper in the journal *Ticks and Tick-Borne Diseases* that contained these unsettling sentences: "Ticks transmit a greater variety of pathogenic, disease-causing agents than any other blood-sucking arthropod," it began, somewhat ominously. "Despite more than a century of efforts, control of ticks and tick-borne diseases remains a daunting challenge throughout the world." Indeed. So far, the challenge of a vaccine had been unmet. The baited boxes for mice, described above, were effective but spotty and costly. Deer were fenced, culled, and killed, to little long-term effect. Feeding stations had even been tried at which deer thrust their heads between sets of vertical rollers in pursuit of a bite of corn and, like mice, were doused with an acaricide. After a five-year test, officials in Fairfax County, Virginia, concluded they did not want to set up diners for deer. Although tick populations declined somewhat, the deer congregated, trampled undergrowth, and, more disturbingly, grew in density. Other studies on the method were at turns promising and disappointing.

If Holly Gaff has her way, little electronic robots may one day troll parks, the edges of ballfields, or wherever people congregate outdoors, all in the service of killing ticks. Her robot is a four-wheeled kind of Superbot, replete with a cape soaked in a chemical that in the case of her experiment was permethrin. The robot even breathed, expelling carbon dioxide in the way that humans do and that makes ticks perk up and extend their forelegs in pursuit of a host. Superbot, or what Gaff calls TickBot, was set loose on three woodland trails in Virginia, its white corduroy cape dragging slowly behind. The results were impressive. Wherever TickBot went, rumbling slightly as an added attractant, the population of lone star ticks was reduced to zero within an hour and stayed at or near there for a full day. Unlike bait boxes and feeding stations, the effect was immediate and thorough. "An hour afterwards, we could sit on the ground in the field site there and not get a tick on us," Gaff told me. "That was crazy. Lone stars will chase you down." For this, she thanks a friend at Virginia Military Institute who got Lyme disease and thought to put a remote-controlled car to novel use.

There are, of course, problems with this approach. The ticks came back after a day. TickBot requires a guide wire to define its course, what Gaff called "a little breadcrumb trail for the robot," similar to invisible fencing used to contain dogs. The device needs its batteries charged and to be overseen, lest it be hijacked from a public park, for example. It also isn't known if it will work on the blacklegged ticks in the eastern US, or *Ixodes* anywhere, that carry Lyme disease. But as with bait boxes and tick-repelling oils, TickBot has potential. Gaff had moved to phase two when I spoke with her, testing it in a child care center and in a neighborhood while substituting a piece of dry ice for piped-in CO_2. The ice achieves the same result—making the tick think it's a mammal. "It's a living, breathing, moving machine now," Gaff said. And it obliterates lone star ticks.

Around the time I spoke with Gaff, New York was experiencing a particularly bad tick season, the result of a bumper crop of acorns two years earlier that had led to more tick-infested mice a year later, and, this year, to many more ticks. We can thank a changing climate for the calamity of large-tree seeds that set this event in motion. This trend had landed a tick in my groin, my best efforts notwithstanding. When I passed fields full of ball-playing youngsters at the local elementary school, I virtually closed my eyes in fear. There they were, playing on a grassy expanse bordered by a thin strip of woods and a fringe of weeds, the places ticks like best. A pre-sweep with Holly Gaff's robot, I thought, might just help protect those children, if only funding to test it were as prolific as the ticks.

But at this school, where parents eagerly watched their offspring from the sidelines, there wasn't a single sign that warned of ticks in the place where it might do the most good. The grounds were sprayed with a natural tick-deterrent, but only after reports of ticks already on children, which is more than many schools do. The district's overseer of facilities was sympathetic. "I find ticks on my children all of the time," Michael Shore told me, "and it scares the heck out of me." But signs might alarm parents, he said, and routine spraying would be "cost prohibitive."

Had my small town in upstate New York had one or two cases of West Nile or Zika virus, which most times, like Lyme, are mild, manageable illnesses, there would have been warning signs, fumigation, and public panic. Yet Lyme disease can be as vicious as those mosquito-borne diseases and is far more common. People living in endemic areas will most likely be exposed to *Borrelia burgdorferi* at some point in their lives, whether they know they were tick-bitten or not.

Consider a study of blood donors in Sweden, published in 2017. When compared with younger groups, men sixty to seventy years old had the highest rates of antibodies to the Lyme spirochete, an indicator of either past or current exposure. Fifty-two percent were positive. The longer you live, it can be concluded, the greater the chance of a brush with Lyme disease. Many people are infected more than once. Yet so far, the war against Lyme disease has largely been waged through proclamation—May is Lyme Disease Awareness Month—and by press release. People have been told to tuck their socks into their pants, which I've observed few do, and check themselves for ticks. Put clothes in a dryer, after gardening or hiking, to kill arachnids. Spray clothing and shoes with permethrin. Use repellents. Clearly, these things have not stemmed a growing wave of infection from a rising tide of ticks. That's not a reason to give up on these measures. It's a call for governments at all levels to step up, from nudging citizens to ward off tick encounters to funding broader antitick research.

Job Hazard

In the Netherlands, Lyme disease has been considered an occupational risk since a judge ruled in 2009 that a police officer had had substantial on-the-job exposure to ticks before contracting debilitating Lyme disease. A study ensued on how to manage the risk, concluding in 2013 that workers wearing pants impregnated with permethrin had significantly fewer tick bites than those in untreated clothing. A study at the University of Rhode Island, similarly, found the chemical created a "hot-foot" effect that caused agitated ticks to bolt from a treated sleeve or pant leg. The chemical is officially classified a "weak carcinogen" by the US Environmental Protection Agency, however, which raised concerns among Dutch workers. In the end, employers were told to provide protective, though not necessarily treated, clothing. In short: Do something.

Permethrin-impregnated clothing has long been used to ward off malaria, scrub typus, leischmaniasis, and Lyme disease among American military personnel, for whom service in tick- and insect-infested areas is often a greater risk than combat. In 1994, a panel of the National Research Council studied the effects of eighteen-hour-per-day exposure to the clothing for ten years and found service members "unlikely to experience adverse health effects." Tom Mather, the Rhode Island "Tick Guy," has concluded the chemical is safe for humans and dogs, though perhaps not bees and fish. In making the risk-benefit calculation of exposure to ticks or to a common, approved insecticide, consider the turn taken in the life of a young Dutch woman who knows the occupational hazards of Lyme disease.

In 2015, Maaike Boere was twenty-two-years old, employed as a social worker, living on her own, and saving money for a wedding. Then she was bitten by a tick on an outdoor field trip with the disabled people she oversaw. She got a round rash but wrote it off as ringworm. She wasn't treated for another year, when she got very sick. Along the way, this heretofore active young woman with many friends and a bright future was told by several doctors to get over it, she wasn't really sick, she needed physical therapy, she had a mental problem. This is the template of many an infuriating Lyme story. One doctor finally believed her, but it was too late. Two three-month rounds of antibiotics—generous to be sure—did not resolve her serious neurological issues.

I met Maaike, twenty-four, at a Lyme disease event in Amsterdam, where she was every inch a vibrant young woman, with blue eyes, long dark-blond hair, and red lipstick. She was thrilled to be interviewed and insisted that she remove her jacket for a photograph, showing wrists prettied with bracelets and a fringe of black lace at the shoulders of her sleeveless shirt. The only inconsistency in this picture of vitality was the blue air mattress on which she was propped, and where, suddenly, her left leg began to spasm, a neurological manifestation of the Lyme pathogen's wrath. The smile faded, replaced by a mixture of fear and embarrassment, as she clasped her thigh until the tremor passed. A wedding postponed, Maaike and her fiancé were pooling money for a visit to a Belgian specialist that had filled her with hope. "When I ask my doctor about Lyme disease, he gets very angry," she said unprompted. "'Not everybody has Lyme disease,' he says."

When utility workers in the Netherlands fix cables buried beneath weeds or landscapers prune bushes, they will have far more protection against ticks than Maaike Boere did, sparing some from her disability and pain. A standard for permethrin-impregnated clothing will soon be proposed for adoption across the European Union. In the United

States, Lyme disease adds $712 million to $1.3 billion to national health costs every year, according to a Johns Hopkins study. That's an estimated $3,000 per patient. Now consider the unmeasured cost of lost earnings and disability payments. Forty percent of patients who report they have chronic Lyme disease were unable to work, according to a study of 3,000 patients by Lymedisease.org and Carnegie Mellon University.

Maaike, for one, had not worked in nearly two years in a land generous for its social welfare but unable and unwilling to provide care for late-stage Lyme disease. A day after meeting Maaike, a Dutch man, who I met on a bike trail where I was flagging for ticks, told me that after he was bitten by a tick, his two physician brothers had a suggestion: Go for tests to Germany. In that country, diagnostics are used for potential tick-borne illness other than the often-unreliable two-tiered testing. Of course, you must be able to pay for them.

The Dutch clothing regulation brings me back to Tom Mather's chip-away-at-the-problem philosophy. The Netherlands' effort to protect workers represents one salvo against ticks, preventing infection, reducing human misery, and educating workers, who then bring the message home. A savvy scientist in Holland named Sip van Wieren tried a different tack. He treated sheep with an insecticide and let them loose in forest patches and small campsites, which are a hugely popular destination for the Dutch. Van Wieren then watched as the sheep collected ticks by the handful in their furry fleece—"sheep mopping," he called it—reducing populations by 70 percent. With sheep a ubiquitous fixture on the flat Dutch landscape, the experiment employed available resources to great effect.

Van Wieren's sheep, Mather's shoe treatments, Gaff's robots—they are each one strike against tick-borne disease. We need more. Warn and empower citizens at every drugstore checkout. Tell parents their kids are potential meals for ticks. Tell schools. Combine the efforts of a thousand advocacy groups into a coordinated strategy. Employ and subsidize tactics known to work. Moreover, and for the long-term health of the planet, recognize our role in creating this epidemic and the forces that propel it.

Silver Bullet?

In April of 2017, a Yale University ecological economist named Eli Fenichel calculated a different cost of Lyme disease. Call it the fear factor, akin to my new vow to avoid walking in my beloved meadow. Fenichel and three other researchers devised a way to correlate the incidence of Lyme disease and the duration of tick season with the amount of time people spent on outdoor activities, as recorded in the American Time Use Survey. In high-Lyme states in the Northeast and Midwest of the United States, it turned out, people went on fewer hikes in the woods or strolls in the park than people in other parts of the country, the disparity pronounced in tick season but not at other times.

In all, residents of Lyme-ridden areas lost more than nine hours a year in outdoor time. That lost experience, Fenichel and colleagues concluded, was worth $2.8 billion to $5 billion a year—one billion missed forays into the natural world multiplied by the $2.74 to $4.91 travel cost of each. This, of course, did not capture the intrinsic value of breathing forest air. But it was one way to "monetize well-being," as he put it, and for a simple reason. "If there were a silver bullet to get rid of Lyme disease," Fenichel told me from his office at Yale, "we should be willing to pay three to five billion dollars for that silver bullet as a society." And so, the paper asked, "How much would people be willing to pay for a Lyme Disease free world?"

David Whitman sells the two-step tick protection technology that I rejected as too costly. For sixteen years, his company, Connecticut Tick Control, has applied the product along the fringes of sporting fields and playgrounds at seven schools in western Connecticut. I was happy to hear this, the vision of unprotected children in my own community so clear. He agreed the cost was high; another school district he had long serviced had recently cut the protection from its budget. "There are municipalities who say we need to protect the kids while they are in our care," he told me. "There's no silver bullet. There's no cheap way out."

For too long, however, the cheap way has long been the chosen way. We, as a society, have done little to mount an effort against ticks, and some have paid the price. A tick-bite on a grassy preschool playground in Kansas City, Kansas, ended the teaching career of Kathy White in 1998, when she was fifty-three-years old. Active in the Lyme Association of Greater Kansas City, White hands out Lyme disease information kits at schools—our chief method of fighting this epidemic—where nurses have reported using the packets' special tweezers to remove ticks from students. White and a small army of people who learned of Lyme the hard way are chipping away at the problem. They cannot wait for science to solve this. Although it would be nice.

In 2016, an MIT researcher named Kevin Esvelt had the idea to tweak the DNA of mice in a way that would make them immune to invasion by *Borrelia burgdorferi*. Ticks would bite, the idea proposed, but the pathogen would not infect the mice, which are the single biggest, though not only, source of the Lyme disease bug in nature. These genetically altered mice would no longer nourish and sustain the Lyme pathogen, breaking the cycle of circulation—tick to mouse and vice versa. This would involve years of work, of course. First, identify mice that make the best Lyme antibodies after vaccination. Then, implant the bit of DNA that ramps up this prized resistance into the eggs of female mice, insuring it will pass to roughly half of their progeny. The altered mice would be released first onto an uninhabited island off of Cape Cod, Massachusetts, then onto Lyme-riddled Nantucket and Martha's Vineyard islands, where the power of ticks to transmit *Borrelia burgdorferi* would hopefully be defused.

"Sculpting Evolution" is the name of the MIT project of which this is part, and Esvelt is aware of its considerable implications for the natural world and the need for transparency and community support. As he told local residents at a public meeting in July 2016, "Life finds a way." He is determined to avoid unforeseen consequences, in particular if and when the idea is applied toward a mainland fix that is more sweeping and permanent. On the offshore islands, the mice would be altered using DNA from other mice. But to assure success in the wider world, DNA from other species would have to be introduced to assure all mice, not just half, received the mutation. In so doing, the DNA of mice would be forever changed.

Beyond this, Esvelt's technologically alluring fix has two other significant hitches. As with human Lyme disease vaccines, it would still allow ticks to proliferate, bite, and, most significantly, deliver other pathogens besides the vanquished Lyme bug. To address that, the MIT group may try another tack, altering DNA in mice in a way that targets tick saliva like European antitick vaccine. Feeding would be blocked or the tick killed. Esvelt's island experiments would likely run into the low tens of millions, he said. The cost will be far greater for trials in the wider world. "Are there enough Lyme cases here," he asked, "to make that worth it?"

Quarantines and Cattle

So what is the value of avoiding tick-borne infection? Consider the Texas-Mexico border. There, an army of agricultural experts is dedicated to the task of stopping cattle ticks from moving north of the Rio Grande and into prime ranchlands of Texas longhorn cattle. The surveillance program is meant to wipe out cattle fever, also called bovine babesiosis. It is aggressive and expensive. Nearly two million acres of land were under permanent and emergency quarantine as of mid-2017, as the ticks crept farther north on the heels of a warming climate. Under the program, cattle are regularly rounded up, sometimes by helicopter, inspected, vaccinated, and periodically dipped into treatment baths for six to nine months. Deer are fed corn laced with a tick-killing chemical or treated to a dousing of permethrin at special feeding stations. This is all done, free of charge to ranchers, to protect not one but two lucrative industries: beef production and hunting on growing preserves populated with game that also spreads the tick. These efforts have been successful. In the meantime, blacklegged and lone star ticks also move north. They run rampant. They inflict damage. But the response is anemic and inconsequential. And these ticks, unlike their counterparts on cattle, harm people. Many, many people.

We are left, then, with a pandemic against which we chip away. Homeowners scatter cotton balls soaked in permethrin so that passing mice may use them to line their nests and be made tick-free. New York State adopts a law to develop a curriculum on Lyme disease prevention, though schools don't have to teach it. Facebook posts remind us to check for ticks. Conservation groups hold classes on garden plants that deter ticks. Lyme groups hand out pamphlets, demand change, and share news stories, like one I read recently of a two-year-old girl named Kenley who died in Indiana days after a tick infected her with Rocky Mountain spotted fever. This is all something.

Aginar Mafra Neto, a chemical ecologist, is testing his own idea, putting tick pheromones and pesticides into tiny black dollops that he spreads on vegetation; ticks lured to what looks a lot like tick excrement face death or debility. Several scientists told me the idea had potential. Said its inventor, in a refrain I'd heard before: "For ticks, there's no funding." If *Ixodes* ticks were agricultural pests, there would be money, he said.

But perhaps this picture is not as bleak as it sounds. For some projects, there is money, except that it is not coming from the government. In 2016, the Cary Institute of Ecosystem Studies in New York State began a $5 million, five-year experiment to bring down Lyme rates in twenty-four neighborhoods with a one-two punch. Properties are treated with a fungicide that kills ticks, while mice are invited into feeding stations for a coating of insecticide. The funding was among $40 million distributed by the Steven & Alexandra Cohen Foundation of Stamford, Connecticut, which, with other nonprofit organizations, has stepped into the Lyme disease void left by the National Institutes of Health and the CDC. Another nonprofit group that is funding significant research, the Bay Area Lyme Foundation in Silicon Valley, California, computed the US government's research support based on the number of afflicted people. For every HIV/AIDS case, the NIH distributed $57,960 in 2015. The per capita West Nile Virus allotment was $7,050. Lyme disease research grants amounted to $133 for each and every case.

For every way out of the mess of tick-borne disease, there are obstacles and drawbacks. It will cause other problems. It will do half the job. It will cost too much. We are fighting an eight-legged menace that emerged in the early part of the Cretaceous period, about 145 million years ago or just after the Jurassic period. It has had a lot of time to evolve. And *Borrelia burgdorferi*, its chief but not sole cargo, has been around for a good portion of that time, taking equal advantage of the benefits of natural selection. It is marvelously adept at changing, chameleon-like, to fool the immune system into thinking it is no longer there. It persists. What is new in the long history of ticks and *Borrelia burgdorferi* are the conditions of the twenty-first century that have made their lives good: the cut-up forests; the profusion of small animals on which ticks feed and in which the pathogen lives; the paucity of animal predators; and the endless, inviting landscape on which to settle, a frontier made possible by a warmer, less challenging world.

If this first epidemic in the era of climate change is to be controlled, three things must occur. First, the pain of tens, maybe hundreds, of thousands of long-term tick-borne disease sufferers must be recognized. Why solve a problem that has barely been acknowledged? Second, health issues must be addressed, including the need for better tests and treatment trials, and an acceptance that the problem is tick-borne disease, not only Lyme disease. Finally, an organized, coordinated effort must be made to tackle the problem of ticks in the environment and the harm they do. Others can suggest how climate change, with its potential to alter life as we know it, should be addressed.

Without this commitment, we will live in a world in which nature is feared or, for the unschooled, dangerous. While mosquito-borne illnesses come in cycles, ticks are constant and forever. There is no waxing or waning, no sign they will collapse or die off. They only grow in number and place and pathogens.

Children should be able to run in a field, their hands brushing the tops of a row of summer grasses. Their mothers should tell them to play outside in the belief it is good for them. Hikers who go to one place should not return to another with a disease that is unrecognized, unaccepted, and even scorned. Lives and careers should be saved from disability and ruin. At the same time, patients should be able to turn to medicine and doctors for help. They should not have to devise a plan, when the wrong tick bites, that relies on their resources, ingenuity, connections, and faith. Parents should not fear losing their children to tick-induced physical or mental illness. They should not fear child welfare authorities who question their Lyme disease choices.

Finally, let's stop calling this Lyme disease. Call it Borreliosis, perhaps, when it's early and clear cut. Label it Tick Infection Syndrome, maybe, when it's not. Then let's admit the obvious. This is an epidemic. It is global and dangerous. It is spreading to new places on earth and affecting places in the human body, the brain for one, in ways that are not fully understood. History teaches us that medicine sometimes clings fiercely to convictions that are ultimately proven wrong. Lyme disease is one such time. Believe this, because ticks are out there. Whether we live in a city, a suburb, or an exurb; in a small town in a valley or a chalet at the top of a high mountain, we all have some occasion—let's hope so, at least—to commune with meadows, trees, sand dunes, and trails. Right now, many of these places in too many countries are havens for tiny, almost invisible, eight-legged creatures waiting for the next meal. On balance, they have power far greater than our own.

Selected References

Adrion, E.R., J. Aucott, K.W. Lemke, et al. "Health Care Costs, Utilization and Patterns of Care Following Lyme Disease." *PLoS One* 10, no. 2 (2015): e0116767.

Alig, R.J. 2007. "U.S. Land-Use Changes Involving Forests: Trends and Projections" *Transactions of the 72nd North American Wildlife and Natural Resources Conference.*

Allan, B.F., F. Keesing, and R.S. Ostfeld. "Effect of Forest Fragmentation on Lyme Disease Risk." *Conservation Biology* 17, no. 1 (2003): 267–72.

Ang, C.W., D.W. Notermans, M. Hommes, et al. "Large Differences between Test Strategies for the Detection of Anti-*Borrelia* Antibodies Are Revealed by Comparing Eight ELISAs and Five Immunoblots." *European Journal of Clinical Microbiology & Infectious Diseases* 30, no. 8 (2011): 1027–32.

Aronowitz, R.A. "The Rise and Fall of the Lyme Disease Vaccines: A Cautionary Tale for Risk Interventions in American Medicine and Public Health." *Milbank Quarterly* 90, no. 2 (2012): 250–77.

Aucott, J.N. "Posttreatment Lyme Disease Syndrome." *Infectious Disease Clinics of North America* 29, no. 2 (2015): 309–23.

Aucott, J.N., C. Morrison, B. Munoz, et al. "Diagnostic Challenges of Early Lyme Disease: Lessons from a Community Case Series." *BMC Infectious Diseases* 9, no. 79 (2009).

Aucott, J.N., A.W. Rebman, L.A. Crowder, et al. "Post-treatment Lyme disease Syndrome Symptomatology and the Impact on Life Functioning: Is

There Something Here?" *Quality of Life Research: An International Journal of Quality of Life Aspects of Treatment, Care and Rehabilitation* 22, no. 1 (2013): 75–84.

Auwaerter, Paul G. "*Editorial Commentary:* Life after Lyme Disease." *Clinical Infectious Diseases* 61, no. 2 (2015): 248–50.

Bacon, R.M., B.J. Biggerstaff, M.E. Schriefer, et al. "Serodiagnosis of Lyme Disease by Kinetic Enzyme-Linked Immunosorbent Assay Using Recombinant VlsE1 or Peptide Antigens of *Borrelia Burgdorferi* Compared with 2-Tiered Testing Using Whole-Cell Lysates." *Journal of Infectious Diseases* 187, no. 8 (2003): 1187–99.

Barbieri, A.M., José M. Venzal, A. Marcili, et al. "*Borrelia Burgdorferi* Sensu Lato Infecting Ticks of the *Ixodes Ricinus* Complex in Uruguay: First Report for the Southern Hemisphere." *Vector-Borne and Zoonotic Diseases (Larchmont, N.Y.)* 13, no. 3 (2013): 147–53.

Barbour, A.G., and D. Fish. "The Biological and Social Phenomenon of Lyme Disease." *Science* 260, no. 5114 (1993): 1610–16.

Barbour, Alan. "Remains of Infection." *Journal of Clinical Investigation* 122, no. 7 (2012): 2344–46.

Barrett, A.W., and S.E. Little. "Vector-Borne Infections in Tornado-Displaced and Owner-Relinquished Dogs in Oklahoma, USA." *Vector-Borne and Zoonotic Diseases (Larchmont, N.Y.)* 16, no. 6 (2016): 428–30.

Barrett, A.W., B.H. Noden, J.M. Gruntmeir, et al. "County Scale Distribution of *Amblyomma Americanum* (Ixodida: Ixodidae) in Oklahoma: Addressing Local Deficits in Tick Maps Based on Passive Reporting." *Journal of Medical Entomology* 52, no. 2 (2015): 269–73.

Basile, R.C., N.H. Yoshinari, E. Mantovani, et al. "Brazilian Borreliosis with Special Emphasis on Humans and Horses." *Brazilian Journal of Microbiology* 48, no. 1 (2017): 167–72.

Berende, A., H., J.M. ter Hofstede, F.J. Vos, et al. "Randomized Trial of Longer-Term Therapy for Symptoms Attributed to Lyme Disease." *New England Journal of Medicine* 374, no. 13 (2016): 1209–20.

Berry, K., J. Bayham, S.R. Meyer, et al. "The Allocation of Time and Risk of Lyme: A Case of Ecosystem Service Income and Substitution Effects." *Environmental and Resource Economics* April (2017): 1–20.

Blanc, F., N. Philippi, B. Cretin, et al. "Lyme Neuroborreliosis and Dementia." *Journal of Alzheimer's Disease* 41, no. 4 (2014): 1087–93.

Bockenstedt, L.K., D.G. Gonzalez, A.M. Haberman, et al. "Spirochete Antigens Persist near Cartilage after Murine Lyme Borreliosis Therapy." *Journal of Clinical Investigation* 122, no. 7 (2012): 2652–60.

Bockenstedt, L.K., J. Mao, E. Hodzic, et al. "Detection of Attenuated, Noninfectious Spirochetes in *Borrelia Burgdorferi*-Infected Mice after Antibiotic Treatment." *Journal of Infectious Diseases* 186, no. 10 (2002): 1430–37.

Bockenstedt, L.K., and J.D. Radolf. "Editorial Commentary: Xenodiagnosis for Posttreatment Lyme Disease Syndrome: Resolving the Conundrum or Adding to It?" *Clinical Infectious Diseases* 58, no. 7 (2014): 946–48.

Borgermans, L., G. Goderis, J. Vandevoorde, et al. "Relevance of Chronic Lyme Disease to Family Medicine as a Complex Multidimensional Chronic Disease Construct: A Systematic Review." *International Journal of Family Medicine* 2014 (2014): 138016.

Bransfield, R. "Suicide and Lyme and Associated Diseases." *Neuropsychiatric Disease and Treatment* 13 (2017): 1575–87.

———"Lyme Disease, Comorbid Tick-Borne Diseases, and Neuropsychiatric Disorders." *Psychiatric Times* 24, no. 14 (2007): 59–62.

Breitschwerdt, E.B. "Did *Bartonella Henselae* Contribute to the Deaths of Two Veterinarians?" *Parasites & Vectors* 8, no. 1 (2015): 317.

Breitschwerdt, E.B., B.C. Hegarty, R.G. Maggi, et al. "*Rickettsia Rickettsii* Transmission by a Lone Star Tick, North Carolina." *Emerging Infectious Diseases* 17, no. 5 (2011): 873–5.

Breitschwerdt, E.B., R.G. Maggi, P. Farmer, et al. "Molecular Evidence of Perinatal Transmission of *Bartonella Vinsonii* Subsp. *Berkhoffii* and *Bartonella Henselae* to a Child." *Journal of Clinical Microbiology* 48, no. 6 (2010): 2289–93.

Breitschwerdt, E.B., R.G. Maggi, W.L. Nicholson, et al. "*Bartonella sp.* Bacteremia in Patients with Neurological and Neurocognitive Dysfunction." *Journal of Clinical Microbiology* 9, no. 46 (2008): 2856–61.

Burdge, D.R., and D.P. O'Hanlon. "Experience at a Referral Center for Patients with Suspected Lyme Disease in an Area of Nonendemicity: First 65 Patients." *Clinical Infectious Diseases* 16, no. 4 (1993): 558–60.

Burgdorfer, W. "Discovery of the Lyme Disease Spirochete and Its Relation to Tick Vectors." *Yale Journal of Biology and Medicine* 57, no. 4 (1984): 515–20.

Burgdorfer, W., R.S. Lane, A.G. Barbour, et al. "The Western Black-Legged Tick, *Ixodes Pacificus*: A Vector of *Borrelia Burgdorferi*." *American Journal of Tropical Medicine and Hygiene* 34, no. 5 (1985): 925–30.

Cairns, V., and J. Godwin. "Post-Lyme Borreliosis Syndrome: A Meta-Analysis of Reported Symptoms." *International Journal of Epidemiology* 34, no. 6 (2005): 1340–45.

Cameron, D.J., L.B. Johnson, and E.L. Maloney. "Evidence Assessments and Guideline Recommendations in Lyme Disease: The Clinical Management of Known Tick Bites, Erythema Migrans Rashes and Persistent Disease." *Expert Review of Anti-Infective Therapy* 12, no. 9 (2014): 1103–35.

Capasso, L. "5300 years ago, the Ice Man Used Natural Laxatives and Antibiotics." *Lancet* 352, no. 9143 (1998): 1864.

Caskey, J.R., and M.E. Embers. "Persister Development by Borrelia burgdorferi Populations In Vitro." *Antimicrobial Agents and Chemotherapy*. 59, no. 10 (2015): 6288-6295.

Centers for Disease Control and Prevention. 2008. "Surveillance for Lyme Disease—United States, 1992–2006." *Morbidity and Mortality Weekly* 2013. "CDC Provides Estimate of Americans Diagnosed with Lyme Disease Each Year." 2013. "Three Sudden Cardiac Deaths Associated with Lyme Carditis—United States, November 2012–July 2013." *Morbidity and Mortality Weekly Report*. 2015. "Lyme Disease: Signs and Symptoms of Untreated Lyme Disease." 2016. "Final 2015 Reports of Nationally Notifiable Infectious Diseases and Conditions." *Morbidity and Mortality Weekly Report*. 2016. "Post-Treatment Lyme Disease Syndrome." 2016. "Natural Tick Repellents and Pesticides." 2016. "Current Guidelines, Common Clinical Pitfalls, and Future Directions for Laboratory Diagnosis of Lyme Disease, United States." 2017. "Antibiotic/Antimicrobial Resistance." 2017. "Research into Prolonged Treatment for Lyme Disease."

Cerar, T., F. Strle, D. Stupica, et al. "Differences in Genotype, Clinical Features, and Inflammatory Potential of *Borrelia Burgdorferi* Sensu Stricto Strains from Europe and the United States." *Emerging Infectious Diseases* 22, no. 5 (2016): 818–27.

Chalada, M.J., J. Stenos, and R.S. Bradbury. "Is There a Lyme-like Disease in Australia? Summary of the Findings to Date." *One Health* 2 (2016): 42–54.

Cheng, A., D. Chen, K. Woodstock, et al. "Analyzing the Potential Risk of Climate Change on Lyme Disease in Eastern Ontario, Canada, Using Time Series Remotely Sensed Temperature Data and Tick Population Modelling." *Remote Sensing* 9, no. 6 (2017): 609.

Chmela, J., J. Kotál, J. Kopecký, et al. "All for One and One for All on the Tick-Host Battlefield." *Trends in Parasitology* 32, no. 5 (2016): 368–77.

Chmielewska-Badora, J., E. Cisak, and J. Dutkiewicz. "Lyme Borreliosis and Multiple Sclerosis: Any Connection? A Seroepidemic Study." *Annals of Agricultural and Environmental Medicine* 7, no. 2 (2000): 141–3.

Clark, K.L., B. Leydet, and S. Hartman. "Lyme Borreliosis in Human

Patients in Florida and Georgia, USA." *International Journal of Medical Sciences* 10, no. 7 (2013): 915–31.

Clarke, C.S., E.T. Rogers, and E.L. Egan. "Babesiosis: Under-Reporting or Case-Clustering?" *Postgraduate Medical Journal* 65, no. 766 (1989): 591–3.

Cohen, E. B., L.D. Auckland, P. P. Marra, and S. A. Hamer. 2015. "Avian Migrants Facilitate Invasions of Neotropical Ticks and Tick-Borne Pathogens into the United States." *Applied and Environmental Microbiology*, October, AEM.02656–15.

Commins, S., S. Satinover, J. Hosen, et al. "Delayed Anaphylaxis, Angioedema, or Urticaria after Consumption of Red Meat in Patients with IgE Antibodies Specific for Galactose-Alpha-1,3-Galactose." *Journal of Allergy and Clinical Immunology* 123, no. 2 (2009): 426–33.

Cook, M., and B. Puri. "Application of Bayesian Decision-Making to Laboratory Testing for Lyme Disease and Comparison with Testing for HIV." *International Journal of General Medicine* 10, April (2017): 113–23.

Cook, M., and B. Puri. "Commercial Test Kits for Detection of Lyme Borreliosis: A Meta-Analysis of Test Accuracy." *International Journal of General Medicine* 9 (2016): 427–40.

Costello, J.M., M.E. Alexander, K.M. Greco, et al. "Lyme Carditis in Children: Presentation, Predictive Factors, and Clinical Course." *Pediatrics* 123, no. 5 (2009): e835–41.

County of Fairfax. 2016. *A Study Report on the Use of 4-Poster Deer Treatment Stations to Control Tick Infestations on White-tailed Deer (Odocoileus virginianus) in Fairfax County, Virginia.* Fairfax, Va.

Crimmins, A., J. Balbus, J.L. Gamble, et al., eds. *The Impacts of Climate Change on Human Health in the United States: A Scientific Assessment.* Washington, DC: U.S. Global Change Research Program, 2016.

Cutler, S.J., E. Ruzic-Sabljic, and A. Potkonjak. "Emerging Borreliae – Expanding beyond Lyme Borreliosis." *Molecular and Cellular Probes* 31 (2017): 22–27.

Dahlgren, F.S., C.D. Paddock, Y.P. Springer, et al. "Expanding Range of *Amblyomma Americanum* and Simultaneous Changes in the Epidemiology of Spotted Fever Group Rickettsiosis in the United States." *American Journal of Tropical Medicine and Hygiene* 94, no. 1 (2016): 35–42.

Dattwyler, R., D.J. Volkman, B.J. Luft, et al. "Seronegative Lyme Disease. Dissociation of Specific T- and B-lymphocyte Responses to *Borrelia Burgdorferi*." *New England Journal of Medicine* 319, no. 22 (1988): 1441–466.

De Groot, M., B. Baker, and E. Bird. *"Practical Test of Tick-resistant Work*

Wear." (Translated from Dutch). Association of Forest Owners and Nature Terrain, 2013.

DeBiasi, R.L. "A Concise Critical Analysis of Serologic Testing for the Diagnosis of Lyme Disease." *Current Infectious Disease Reports* 16, no. 12 (2014): 450.

Dell'Amore, C. "What's a Ghost Moose? How Ticks Are Killing an Iconic Animal." *National Geographic* June (2015): 1.

DeLong, A.K., B. Blossom, E. Maloney, et al. "Antibiotic Retreatment of Lyme Disease in Patients with Persistent Symptoms: A Biostatistical Review of Randomized, Placebo-Controlled, Clinical Trials." *Contemporary Clinical Trials* 33, no. 6 (2012): 1132–42.

Department of Defense. "Lyme Disease among U.S. Military Members, Active and Reserve Component, 2001–2008." *Medical Surveillance Monthly Report* 16, no. 7 (2009): 2–4.

Dersch, R., T. Hottenrott, S. Schmidt, et al. "Efficacy and Safety of Pharmacological Treatments for Lyme Neuroborreliosis in Children: A Systematic Review." *BMC Neurology* 16, no. 1 (2016): 189.

Dersch, R., H. Sommer, S. Rauer, et al. "Prevalence and Spectrum of Residual Symptoms in Lyme Neuroborreliosis after Pharmacological Treatment: A Systematic Review." *Journal of Neurology* 263, no. 1 (2015): 17–24.

Diuk-Wasser, M.A., E. Vannier, and P.J. Krause. "Coinfection by *Ixodes* Tick-Borne Pathogens: Ecological, Epidemiological, and Clinical Consequences." *Trends in Parasitology* 32, no. 1 (2016): 30–42.

Donta, S.T. "Tetracycline Therapy for Chronic Lyme Disease." *Clinical Infectious Diseases* 25, no. s1, Suppl. 1 (1997): S52–6.

Donta, S.T., R.B. Noto, and J.A. Vento. "SPECT Brain Imaging in Chronic Lyme Disease." *Clinical Nuclear Medicine* 37, no. 9 (2012): e219–22.

Dressler, F., J.A. Whalen, B.N. Reinhardt, et al. "Western Blotting in the Serodiagnosis of Lyme Disease." *Journal of Infectious Diseases* 167, no. 2 (1993): 392–400..

Dunn, J.M., P.J. Krause, S. Davis, et al. "*Borrelia Burgdorferi* Promotes the Establishment of *Babesia Microti* in the Northeastern United States." *PLoS One* 9, no. 12 (2014): e115494.

Ebady, R., A.F. Niddam, A.E. Boczula, et al. "Biomechanics of *Borrelia Burgdorferi* Vascular Interactions." *Cell Reports* 16, no. 10 (2016): 2593–604.

Ehrlich, G.D., R. Veeh, X. Wang, et al. "Mucosal Biofilm Formation on Middle-Ear Mucosa in the Chinchilla Model of Otitis Media." *Journal of the American Medical Association* 287, no. 13 (2002): 1710–15.

Eisen, L., D. Rose, R. Prose, N. Breuner, M. Dolan, K. Thompson, and N. Connally. 2017. "Bioassays to Evaluate Non-Contact Spatial Repellency, Contact Irritancy, and Acute Toxicity of Permethrin-Treated Clothing against Nymphal *Ixodes Scapularis* Ticks." *Ticks and Tick-borne Diseases,* October:837–49.

Eisen, R.J., L. Eisen, and C.B. Beard. "County-Scale Distribution of *Ixodes scapularis* and *Ixodes pacificus* (Acari: Ixodidae) in the Continental United States." *Journal of Medical Entomology* 53, no. 2 (2016): 349–86.

Elbaum-Garfinkle, S. "Close to Home: A History of Yale and Lyme Disease." *Yale Journal of Biology and Medicine* 84, no. 2 (2011): 103–8.

Elliott, D.J., S.C. Eppes, and J.D. Klein. "Teratogen Update: Lyme Disease." *Teratology* 64, no. 5 (2001): 276–81.

Elsner, R.A., C.J. Hastey, K.J. Olsen, et al. "Suppression of Long-Lived Humoral Immunity Following *Borrelia Burgdorferi* Infection." *PLoS Pathogens* 11, no. 7 (2015): e1004976.

Embers, M.E., S.W. Barthold, J.T. Borda, et al. "Persistence of *Borrelia Burgdorferi* in Rhesus Macaques Following Antibiotic Treatment of Disseminated Infection." *PLoS One* 7, no. 1 (2012): e29914.

Embers, M.E., and S. Narasimhan. "Vaccination against Lyme Disease: Past, Present, and Future." *Frontiers in Cellular and Infection Microbiology* 3, no. 6 (2013).

Environmental Protection Agency, *Climate Change Indicators in the United States,* 2014, 3rd ed." EPA 430-R-14-004.

Fallon, B.A., J.G. Keilp, K.M. Corbera, et al. "A Randomized, Placebo-Controlled Trial of Repeated IV Antibiotic Therapy for Lyme Encephalopathy." *Neurology* 70, no. 13 (2008): 992–1003.

Fallon, B.A., J. Keilp, I. Prohovnik, et al. "Regional Cerebral Blood Flow and Cognitive Deficits in Chronic Lyme Disease." *Journal of Neuropsychiatry and Clinical Neurosciences* 15, no. 3 (2003): 326–32.

Fallon, B.A., J.M. Kochevar, A. Gaito, et al. "The Underdiagnosis of Neuropsychiatric Lyme Disease in Children and Adults." *Psychiatric Clinics of North America* 21, no. 3 (1998): 693–703.

Fallon, B.A., and J.A. Nields. "Lyme Disease: A Neuropsychiatric Illness." *American Journal of Psychiatry* 151, no. 11 (1994): 1571–83.

Fallon, B.A., E. Petkova, J.G. Keilp, et al. "A Reappraisal of the U.S. Clinical Trials of Post-Treatment Lyme Disease Syndrome." *Open Neurology Journal* 6, Suppl 1 (2012): 79–87.

Fang, D.C., and J. McCullough. "Transfusion-Transmitted *Babesia Microti.*" *Transfusion Medicine Reviews* 30, no. 3 (2016): 132–38.

Feder, H.M., Jr., B.J.B. Johnson, S. O'Connell, et al., and the Ad Hoc International Lyme Disease Group. "A Critical Appraisal of 'Chronic Lyme Disease.'" *New England Journal of Medicine* 357, no. 14 (2007): 1422–30.

Feng, J., P.G. Auwaerter, and Y. Zhang. "Drug Combinations against *Borrelia Burgdorferi* Persisters *In Vitro*: Eradication Achieved by Using Daptomycin, Cefoperazone and Doxycycline." *PLoS One* 10, no. 3 (2015): e0117207.

Feng, J., W. Shi, S. Zhang, et al. "A Drug Combination Screen Identifies Drugs Active against Amoxicillin-Induced Round Bodies of *In Vitro Borrelia Burgdorferi* Persisters from an FDA Drug Library." *Frontiers in Microbiology* 7 (2016): 743.

Feng, J., T. Wang, W. Shi, et al. "Identification of Novel Activity against *Borrelia Burgdorferi* Persisters Using an FDA Approved Drug Library." *Emerging Microbes & Infections* 3, no. 7 (2014): e49.

Feng, J., S. Zhang, W. Shi, et al. "Activity of Sulfa Drugs and Their Combinations against Stationary Phase B. burgdorferi In Vitro." *Antibiotics (Basel, Switzerland)* 6, no. 1 (2017): 10.

Fife, C.E., and K.A. Eckert. "Hyperbaric Oxygen Therapy and Chronic Lyme Disease: The Controversy and the Evidence." In *Textbook of Hyperbaric Medicine*, ed. Kewal K. Jain, 171–81. Switzerland: Springer, 2016.

Fraser, C.M., S. Casjens, W.M. Huang, et al. "Genomic Sequence of a Lyme Disease Spirochaete, *Borrelia Burgdorferi*." *Nature* 390, no. 6660 (1997): 580–86.

Gaff, H.D., A. White, K. Leas, et al. "TickBot: A Novel Robotic Device for Controlling Tick Populations in the Natural Environment." *Ticks and Tick-Borne Diseases* 6, no. 2 (2015): 146–51.

Gardner, T. "Lyme Disease." In *Infectious Diseases of the Fetus and Newborn Infant*, ed. J.S. Remington and J.O. Klein, 519–642. Philadelphia: Saunders, 2001.

Ginsberg, H.S., M. Albert, L. Acevedo, et al. "Environmental Factors Affecting Survival of Immature *Ixodes Scapularis* and Implications for Geographical Distribution of Lyme Disease: The Climate/Behavior Hypothesis." *PLoS One* 12, no. 1 (2017): e0168723.

Godfrey, E.R., and S.E. Randolph. "Economic Downturn Results in Tick-borne Disease Upsurge." *Parasites & Vectors* 4, no. 1 (2011): 35.

Government of Canada. "National Lyme Disease Surveillance in Canada 2013: Web Report."

Grab, D.J., G. Perides, J.S. Dumler, et al. "*Borrelia Burgdorferi*, Host-Derived Proteases, and the Blood-Brain Barrier." *Infection and Immunity* 73, no. 2 (2005): 1014–22.

Grann, D. 2001. "Stalking Dr. Steere." *The New York Times Magazine,* June 17.

Granter, S.R., R.S. Ostfeld, and D.A. Milner Jr. "Where the Wild Things Aren't: Loss of Biodiversity, Emerging Infectious Diseases, and Implications for Diagnosticians." *American Journal of Clinical Pathology* 146, no. 6 (2016): 644–6.

Gray, J.S., H. Dautel, A. Estrada-Peña, et al. "Effects of Climate Change on Ticks and Tick-Borne Diseases in Europe." *Interdisciplinary Perspectives on Infectious Diseases* 2009 (2009): 1–12.

Habeggar, S. 2014. "Lyme Disease in Canada: An Update on the Epidemiology." *Purple Paper* (43).

Hájek, T., B. Pašková, D. Janovská, et al. "Higher Prevalence of Antibodies to *Borrelia Burgdorferi* in Psychiatric Patients than in Healthy Subjects." *American Journal of Psychiatry* 159, no. 2 (2002): 297–301.

Han, S., G.J. Hickling, and J.I. Tsao. "High Prevalence *of Borrelia miyamotoi* among Adult Blacklegged Ticks from White-Tailed Deer." *Emerging Infectious Diseases* 22, no. 2 (2016): 316–18.

Hao, Q., X. Hou, Z. Geng, et al. "Distribution of *Borrelia Burgdorferi* Sensu Lato in China." *Journal of Clinical Microbiology* 49, no. 2 (2011): 647–50.

Hasle, G. "Transport of Ixodid Ticks and Tick-Borne Pathogens by Migratory Birds." *Frontiers in Cellular and Infection Microbiology* 3, September (2013): 1–6.

Hermance, M.E., and S. Thangamani. "Powassan Virus: An Emerging Arbovirus of Public Health Concern in North America." *Vector-Borne and Zoonotic Diseases (Larchmont, N.Y.)* 17, no. 7 (2017): 453–62.

Herman-Giddens, M.E. "Tick-Borne Diseases in the South-East Need Human Studies: Lyme Disease, STARI and Beyond." *Zoonoses and Public Health* 61, no. 1 (2014): 1–3.

Herrmann, C., and L. Gern. "Search for Blood or Water Is Influenced by *Borrelia Burgdorferi* in *Ixodes Ricincus Parasites & Vectors* 8, no. 1 (2015): 6.

Hersh, M.H., R.S. Ostfeld, D.J. McHenry, et al. "Co-Infection of Black-legged Ticks with *Babesia Microti* and *Borrelia Burgdorferi* Is Higher than Expected and Acquired from Small Mammal Hosts." *PLoS One* 9, no. 6 (2014): e99348.

Herwaldt, B.L., J.V. Linden, E. Bosserman, et al. "Transfusion-Associated Babesiosis in the United States: A Description of Cases." *Annals of Internal Medicine* 155, no. 8 (2011): 509–19.

Hodzic, E., D. Imai, S. Feng, et al. "Resurgence of Persisting Non-Cultivable

Borrelia Burgdorferi following Antibiotic Treatment in Mice." *PLoS One* 9, no. 1 (2014): e86907.

Hofhuis, A., M. Harms, C. van den Wijngaard, et al. "Continuing Increase of Tick Bites and Lyme Disease between 1994 and 2009." *Ticks and Tick-Borne Diseases* 6, no. 1 (2015): 69–74.

Hofhuis, A., T. Herremans, D.W. Notermans, et al. "A Prospective Study among Patients Presenting at the General Practitioner with a Tick Bite or Erythema Migrans in the Netherlands." *PLoS One* 8, no. 5 (2013): e64361.

Holzbauer, S.M., M.M. Kemperman, and R. Lynfield. "Death Due to Community-Associated *Clostridium Difficile* in a Woman Receiving Prolonged Antibiotic Therapy for Suspected Lyme Disease." *Clinical Infectious Diseases* 51, no. 3 (2010): 369–70.

Horobik, V., F. Keesing, and R.S. Ostfeld. "Abundance and *Borrelia Burgdorferi*-Infection Prevalence of Nymphal *Ixodes Scapularis* Ticks along Forest-Field Edges." *EcoHealth* 3, no. 4 (2006): 262–68.

Horowitz, R.I. 1998. "Atovaquone and Azithromycin Therapy: A New Treatment Protocol for Babesiosis in Co-infected Lyme Patients." Abstract, *International Scientific Conference on Lyme Disease and Other Spirochetal & Tick-borne Disorders*, New York, N.Y.

Horowitz, R., and P. R. Freeman. 2016. "The Use of Dapsone as a Novel 'Persister' Drug in the Treatment of Chronic Lyme Disease/Post Treatment Lyme Disease Syndrome." *Clinical & Experimental Dermatology Research* 7 (3).

Humphrey, P.T., D.A. Caporale, and D. Brisson. "Uncoordinated Phylogeography of *Borrelia Burgdorferi* and Its Tick Vector, *Ixodes Scapularis*." *Evolution* 64, no. 9 (2010): 2653–63.

Hyde, J.A. "*Borrelia Burgdorferi* Keeps Moving and Carries on: A Review of Borrelial Dissemination and Invasion." *Frontiers in Immunology* 8 (2017): 114.

Institute of Medicine. *Critical Needs and Gaps in Understanding Prevention, Amelioration, and Resolution of Lyme and Other Tick-Borne Diseases.* National Academies Press, 2011.

Ivanova, L.B., A. Tomova, D. González-Acuña, et al. "*Borrelia Chilensis*, a New Member of the *Borrelia Burgdorferi* Sensu Lato Complex That Extends the Range of This Genospecies in the Southern Hemisphere." *Environmental Microbiology* 16, no. 4 (2014): 1069–80.

Jaenson, T.G.T., D.G.E. Jaenson, L. Eisen, et al. "Changes in the Geographical Distribution and Abundance of the Tick *Ixodes Ricinus* during the Past 30 Years in Sweden." *Parasites & Vectors* 5, no. 8 (2012):1–15.

Jobling, M.A. "The Iceman Cometh." *Investigative Genetics* 3, no. 1 (2012): 8.

Johnson, B.J.B. "Laboratory Diagnostic Testing for *Borrelia Burgdorferi* Infection." In *Lyme Disease: An Evidence-based Approach*, ed. John J. Halperin, 73–88., 2011. Oxford,UK: CABI.

Johnson, B.J.B., M.A. Pilgard, and T.M. Russell. "Assessment of New Culture Method for Detection of *Borrelia* Species from Serum of Lyme Disease Patients." *Journal of Clinical Microbiology* 52, no. 3 (2014): 721–24.

Johnson, L., A. Aylward, and R.B. Stricker. "Healthcare Access and Burden of Care for Patients with Lyme Disease: A Large United States Survey." *Health Policy (Amsterdam)* 102, no. 1 (2011): 64–71.

Johnson, L., and R.B. Stricker. "The Infectious Diseases Society of America Lyme guidelines: a cautionary tale about the development of clinical practice guidelines." *Philosophy, Ethics, and Humanities in Medicine.* 5, no. 9 (2010): 1-17.

Johnson, T.L., C.B. Graham, K.A. Boegler, et al. "Prevalence and Diversity of Tick-Borne Pathogens in Nymphal *Ixodes Scapularis* (Acari: Ixodidae) in Eastern National Parks." *Journal of Medical Entomology* 54, no. 3 (2017): 742–51.

Johnson, L., S. Wilcox, J. Mankoff, et al. "Severity of Chronic Lyme Disease Compared to Other Chronic Conditions: A Quality of Life Survey." *PeerJ* 2, March (2014): 1–21.

Jones, K.E., N.G. Patel, M.A. Levy, et al. "Global Trends in Emerging Infectious Diseases." *Nature* 451, no. 7181 (2008): 990–93.

Jore, S., H. Viljugrein, M. Hofshagen, et al. "Multi-source Analysis Reveals Latitudinal and Altitudinal Shifts in Range of *Ixodes Ricinus* at Its Northern Distribution Limit." *Parasites & Vectors* 4, no. 84 (2011).

Joseph, J.T., S.S. Roy, N. Shams, et al. "Babesiosis in Lower Hudson Valley, New York, USA." *Emerging Infectious Diseases* 17, no. 5 (2011): 843–7.

Kazimírová, M., and I. Štibrániová. "Tick Salivary Compounds: Their Role in Modulation of Host Defences and Pathogen Transmission." *Frontiers in Cellular and Infection Microbiology* 3 (2013): 43.

Khatchikian, C.E., M.A. Prusinski, M. Stone, et al. "Recent and Rapid Population Growth and Range Expansion of the Lyme Disease Tick Vector, *Ixodes Scapularis*, in North America." *Evolution* 69, no. 7 (2015): 1678–89.

Kirkland, K.B., T.B. Klimko, R.A. Meriwether, et al. "Erythema Migrans-like Rash Illness at a Camp in North Carolina: A New Tick-Borne Disease?" *Archives of Internal Medicine* 157, no. 22 (1997): 2635–41.

Klempner, M.S., L.T. Hu, J. Evans, et al. "Two Controlled Trials of Antibiotic Treatment in Patients with Persistent Symptoms and a History of Lyme Disease." *New England Journal of Medicine* 345, no. 2 (2001): 85–92.

Knox, K.K., A.M. Thomm, Y.A. Harrington, et al. "Powassan/Deer Tick Virus and *Borrelia Burgdorferi* Infection in Wisconsin Tick Populations." *Vector-Borne and Zoonotic Diseases (Larchmont, N.Y.)* 17, no. 7 (2017): 463–66.

Kolata, Gina. 2001. "Lyme Disease Is Hard to Catch And Easy to Halt, Study Finds." *The New York Times*, June 13.

Krause, P.J., S.R. Telford, A. Spielman, et al. "Concurrent Lyme Disease and Babesiosis: Evidence for Increased Severity and Duration of Illness." *Journal of the American Medical Association* 275, no. 21 (1996): 1657–60.

Krupp, L.B., L.G. Hyman, R. Grimson, et al. "Study and Treatment of Post Lyme Disease: A Randomized Double Masked Clinical Trial." *Neurology* 60, no. 12 (2003): 1923–30.

Kugeler, K.J., K.S. Griffith, L.H. Gould, et al. "A Review of Death Certificates Listing Lyme Disease as a Cause of Death in the United States." *Clinical Infectious Diseases* 52, no. 3 (2011): 364–7.

Lancaster, E. "The Diagnosis and Treatment of Autoimmune Encephalitis." *Journal of Clinical Neurology (Seoul, Korea)* 12, no. 1 (2016): 1–13.

Lantos, P.M., R.G. Maggi, B. Ferguson, et al. "Detection of *Bartonella* Species in the Blood of Veterinarians and Veterinary Technicians: A Newly Recognized Occupational Hazard?" *Vector-Borne and Zoonotic Diseases (Larchmont, N.Y.)* 14, no. 8 (2014): 563–70.

Lantos, P.M., and G.P. Wormser. "Chronic Coinfections in Patients Diagnosed with Chronic Lyme Disease: A Systematic Review." *American Journal of Medicine* 127, no. 11 (2014): 1105–10.

Lathrop, S.L., R. Ball, P. Haber, et al. "Adverse Event Reports Following Vaccination for Lyme Disease: December 1998–July 2000." *Vaccine* 20, no. 11–12 (2002): 1603–08.

Lee, S.H. "Lyme Disease Caused by *Borrelia Burgdorferi* with Two Homeologous 16S rRNA Genes: A Case Report." *International Medical Case Reports Journal* 9, no. April (2016): 101–6.

Lee, S.H., J.S. Vigliotti, V.S. Vigliotti, et al. "DNA Sequencing Diagnosis of Off-Season Spirochetemia with Low Bacterial Density in *Borrelia Burgdorferi* and *Borrelia Miyamotoi* Infections." *International Journal of Molecular Sciences* 15, no. 7 (2014): 11364–86.

Leeflang, M.M.G., C.W. Ang, J. Berkhout, et al. "The Diagnostic Accuracy of Serological Tests for Lyme Borreliosis in Europe: A Systematic Review and Meta-Analysis." *BMC Infectious Diseases* 16, no. 140 (2016): 1–17.

Lettau, L.A. "From the Centers for Fatigue Control (CFC) Weekly Report: Epidemiologic Notes and Reports." *Annals of Internal Medicine* 114, no. 7 (1991): 602.

Levi, T., F. Keesing, K. Oggenfuss, et al. "Accelerated Phenology of Black-legged Ticks under Climate Warming." *Philosophical Transactions of the Royal Society of London. Series B, Biological Sciences* 370, no. 1665 (2015): 20130556.

Levi, T., A.M. Kilpatrick, M. Mangel, et al. "Deer, Predators, and the Emergence of Lyme Disease." *Proceedings of the National Academy of Sciences of the United States of America* 109, no. 27 (2012): 10942–47.

Lindgren, E., and T.G.T. Jaenson. *Lyme Borreliosis in Europe: Influences of Climate and Climate Change, Epidemiology, Ecology and Adaption Measures*. World Health Organization, 2006.

Liu, Q., L. Cao, and X. Zhu. "Major Emerging and Re-emerging Zoonoses in China: A Matter of Global Health and Socioeconomic Development for 1.3 billion." *International Journal of Infectious Diseases* 25 (2014): 65–72.

Lyme Disease Association of Australia. 2016. "A Patient Perspective." 2016. *Parliament of Australia Inquiry: Growing Evidence of an Emerging Tick-Borne Disease that Causes a Lyme Like Illness for Many Australian Patients.*

Magnaval, J., I. Leparc-Goffart, M. Gibert, et al. "A Serological Survey about Zoonoses in the Verkhoyansk Area, Northeastern Siberia (Sakha Republic, Russian Federation)." *Vector-Borne and Zoonotic Diseases (Larchmont, N.Y.)* 16, no. 2 (2016): 103–9.

Makhani, N., S.K. Morris, A.V. Page, et al. "A Twist on Lyme: The Challenge of Diagnosing European Lyme Neuroborreliosis." *Journal of Clinical Microbiology* 49, no. 1 (2011): 455–7.

Marques, A. "Chronic Lyme Disease: An Appraisal." *Infectious Disease Clinics of North America* 22, no. 2 (2008): 341–60.

Marques, A., S.R. Telford, III, S. Turk, et al. "Xenodiagnosis to Detect Borrelia Burgdorferi Infection: A First-in-Human Study." *Clinical Infectious Diseases* 58, no. 7 (2014): 937–45.

Marra, P.P., C.M. Francis, R.S. Mulvihill, et al. "The Influence of Climate on the Timing and Rate of Spring Bird Migration." *Oecologia* 142, no. 2 (2005): 307–15.

Marshall, W.F., III, S.R. Telford, III, P.N. Rys, et al. "Detection Of *Borrelia Burgdorferi* DNA In Museum Specimens Of *Peromyscus*." *Journal of Infectious Diseases* 170, no. 4 (1994): 1027–32.

Martinot, M., M.M. Zadeh, Y. Hansmann, et al. "Babesiosis in Immunocompetent Patients, Europe." *Emerging Infectious Diseases* 17, no. 1 (2011): 114–16.

Marzec, N.S., C. Nelson, P.R. Waldron, et al. "Serious Bacterial Infections

Acquired during Treatment of Patients Given a Diagnosis of Chronic Lyme Disease—United States." *Morbidity and Mortality Weekly Report* 66, no. 23 (2017): 607–9.

Materna, J., M. Daniel, and V. Danielová. "Altitudinal Distribution Limit of the Tick *Ixodes Ricinus* Shifted Considerably towards Higher Altitudes in Central Europe: Results of Three Years Monitoring in the Krkonoše Mts. (Czech Republic)." *Central European Journal of Public Health* 13, no. 1 (2005): 24–28.

Matuschka, F., A. Ohlenbusch, H. Eiffert, et al. "Characteristics of Lyme disease Spirochetes in Archived European Ticks." *Journal of Infectious Diseases* 174, no. 2 (1996): 424–6.

McMichael, A.J. "Insights from Past Millennia into Climatic Impacts on Human Health and Survival." *Proceedings of the National Academy of Sciences of the United States of America* 109, no. 13 (2012): 4730–7.

Medlock, J.M., K.M. Hansford, A. Bormane, et al. "Driving Forces for Changes in Geographical Distribution of *Ixodes Ricinus* Ticks in Europe." *Parasites & Vectors* 6, no. 1 (2013): 1–11.

Miklossy JM, K. Khalili, L. Gern, et al. "*Borrelia Burgdorferi* Persists in the Brain in Chronic Lyme Neuroborreliosis and May Be Associated with Alzheimer's Disease." *Journal of Alzheimer's Disease* 6, no. 6 (2004): 639–49.

Mills, L.S., M. Zimova, J. Oyler, et al. "Camouflage Mismatch in Seasonal Coat Color due to Decreased Snow Duration." *Proceedings of the National Academy of Sciences of the United States of America* 110, no. 18 (2013): 7360–5.

Moritz, E.D., C.S. Winton, L. Tonnetti, et al. "Screening for *Babesia Microti* in the U.S. Blood Supply." *New England Journal of Medicine* 375, no. 23 (2016): 2236–45.

Morshed, M.G., J.D. Scott, K. Fernando, et al. "Migratory Songbirds Disperse Ticks across Canada, and First Isolation of the Lyme Disease Spirochete, *Borrelia Burgdorferi*, from the Avian Tick, *Ixodes Auritulus*." *Journal of Parasitology* 91, no. 4 (2005): 780–90.

Moutailler, S., C.V. Moro, E. Vaumourin, et al. "Co-Infection of Ticks: The Rule Rather Than the Exception." *PLoS Neglected Tropical Diseases* 10, no. 3 (2016): e0004539.

Nadelman, R.B., K. Hanincová, P. Mukherjee, et al. "Differentiation of Reinfection from Relapse in Recurrent Lyme Disease." *New England Journal of Medicine* 367, no. 20 (2012): 1883–90.

Nadelman, R.B., J. Nowakowski, D. Fish, et al. "Prophylaxis with Single-Dose

Doxycycline for the Prevention of Lyme Disease after an *Ixodes S capularis* Tick Bite." *New England Journal of Medicine* 345, no. 2 (2001): 79–84.

National Research Council. *Health Effects of Permethrin-Impregnated Army Battle-Dress Uniforms.* Washington, DC: The National Academies Press, 1994.

Nelson, C.A., S. Saha, K.J. Kugeler, et al. "Incidence of Clinician-Diagnosed Lyme Disease, United States, 2005–2010." *Emerging Infectious Diseases* 21, no. 9 (2015): 1625–31.

New York State Department of Environmental Conservation. 2017. "Impacts of Climate Change in New York."

Ogden, N.H., L.R. Lindsay, K. Hanincová, et al. "Role of Migratory Birds in Introduction and Range Expansion of *Ixodes Scapularis* Ticks and of *Borrelia Burgdorferi* and *Anaplasma Phagocytophilum* in Canada." *Applied and Environmental Microbiology* 74, no. 6 (2008): 1780–90.

Ogden, N.H., L.R. Lindsay, and P.A. Leighton. "Predicting the Rate of Invasion of the Agent of Lyme Disease *Borrelia Burgdorferi.*" *Journal of Applied Ecology* 50, no. 2 (2013): 510–18.

Ogden, N. H., G. Margos, D. M. Aanensen, et al. "Investigation of Genotypes of *Borrelia Burgdorferi* in *Ixodes Scapularis* Ticks Collected during Surveillance in Canada." *Applied and Environmental Microbiology* 77, no. 10 (2011): 3244–54.

Ogden, N.H., L.R. Lindsay, M. Morshed, et al. "The Emergence of Lyme Disease in Canada." *Canadian Medical Association Journal* 180, no. 12 (2009): 1221–4.

Ogden, N.H., L. St-Onge, I.K. Barker, et al. "Risk Maps for Range Expansion of the Lyme Disease Vector, *Ixodes Scapularis,* in Canada Now and with Climate Change." *International Journal of Health Geographics* 7, May (2008): 24.

Omeragic, J. "Ixodid Ticks in Bosnia and Herzegovina." *Experimental & Applied Acarology* 53, no. 3 (2011): 301–9.

Ostfeld, R.S., and F. Keesing. "Biodiversity and Disease Risk: the Case of Lyme Disease." *Conservation Biology* 14, no. 3 (2000): 722–8.

Ostfeld, R.S. "A Candid Response to Panglossian Accusations by Randolph and Dobson: Biodiversity Buffers Disease." *Parasitology* 140, no. 10 (2013): 1196–8.

———*Lyme Disease: The Ecology of a Complex System.* Oxford, UK: Oxford University Press, 2011.

Ostfeld, R.S., C.D. Canham, K. Oggenfuss, et al. "Climate, Deer, Rodents, and Acorns as Determinants of Variation in Lyme-Disease Risk." *PLoS Biology* 4, no. 6 (2006): e145.

Ostfeld, R.S., A. Price, V.L. Hornbostel, et al. "Controlling Ticks and Tick-borne Zoonoses with Biological and Chemical Agents." *Bioscience* 56, no. 5 (2006): 383–94.

Oswalt, S.N., and W.B. Smith, eds. *U.S. Forest Resource Facts and Historical Trends.* United States Department of Agriculture, 2014.

Owen, J. "5 Surprising Facts about Otzi the Iceman." *National Geographic* October (2013): 18.

Patel, R., K.L. Grogg, W.D. Edwards, et al. "Death from Inappropriate Therapy for Lyme Disease." *Clinical Infectious Diseases* 31, no. 4 (2000): 1107–9.

PBL Netherlands Environmental Assessment Agency. "The Effects of Climate Change in the Netherlands: 2012."

Pérez de León, A.A., D.A. Strickman, D.P. Knowles, et al. "One Health Approach to Identify Research Needs in Bovine and Human Babysioses: Workshop Report." *Parasites & Vectors* 3, no. 1 (2010): 36.

Persing, D.H., S.R. Telford, III, P.N. Rys, et al. "Detection of *Borrelia Burgdorferi* DNA in Museum Specimens of *Ixodes Dammini* Ticks." *Science* 249, no. 4975 (1990): 1420–3.

Poinar, G., Jr. "Spirochete-like Cells in a Dominican Amber *Ambylomma* Tick (Arachnida: Ixodidae)." *Historical Biology* 27, no. 5 (2015): 565–70.

———"Rickettsial-like Cells in the Cretaceous Tick, *Cornupalpatum Burmanicum* (Ixodida: Ixodidae." *Cretaceous Research* 52. (2014): 1–5.

Ramamoorthi, N., S. Narasimhan, U. Pal, et al. "The Lyme Disease Agent Exploits a Tick Protein to Infect the Mammalian Host." *Nature* 436, no. 7050 (2005): 573–37.

Randolph, S.E., and A.D.M. Dobson. "Pangloss Revisited: A Critique of the Dilution Effect and the Biodiversity-Buffers-Disease Paradigm." *Parasitology* 139, no. 7 (2012): 847–63.

Regier, Y., F. O'Rourke, and V. A. J. Kempf. "*Bartonella* spp: A Chance to Establish One Health Concept in Veterinary and Human Medicine." *Parasites & Vectors* 9, no. 1 (2016): 261.

Reid, M.C., R.T. Schoen, J. Evans, et al. "The Consequences of Overdiagnosis and Overtreatment of Lyme Disease: An Observational Study." *Annals of Internal Medicine* 128, no. 5 (1998): 354–62.

Reik, L., Jr., W. Burgdorfer, and J.O. Donaldson. "Neurologic Abnormalities in Lyme Disease without Erythema Chronicum Migrans." *American Journal of Medicine* 81, no. 1 (1986): 73–78.

Richer, L.M., D. Brisson, R. Melo, et al. "Reservoir Targeted Vaccine against *Borrelia Burgdorferi*: A New Strategy to Prevent Lyme Disease Transmission." *Journal of Infectious Diseases* 209, no. 12 (2014): 1972–80.

Richter, D., F.M. Matuschka, A. Spielman, et al. "2013. "How Ticks Get under Your Skin: Insertion Mechanics of the Feeding Apparatus of *Ixodes Ricinus* Ticks." *Proceedings of the Royal Society B: Biological Sciences* 280 (1773): 1758.

Rizzoli, A., H.C. Hauffe, G. Carpi, et al. "Lyme Borreliosis in Europe." *Eurosurveillance* 16, no. 27 (2011): 1–8.

Rollend, L., D. Fish, and J.E. Childs. "Transovarial Transmission of *Borrelia* Spirochetes by *Ixodes Scapularis*: A Summary of the Literature and Recent Observations." *Ticks and Tick-Borne Diseases* 4, no. 1–2 (2013): 46–51.

Rose, C.D., P.T. Fawcett, K.M. Gibney, et al. "The Overdiagnosis of Lyme Disease in Children Residing in an Endemic Area." *Clinical Pediatrics* 33, no. 11 (1994): 663–68.

Roy-Dufresne, E., T. Logan, J.A. Simon, et al. "Poleward Expansion of the White-Footed Mouse (*Peromyscus leucopus*) under Climate Change: Implications for the Spread of Lyme Disease." *PLoS One* 8, no. 11 (2013): e80724.

Rudenko, N., M. Golovchenko, M. Vancova, et al. "Isolation of Live *Borrelia Burgdorferi* Sensu Lato Spirochaetes from Patients with Undefined Disorders and Symptoms Not Typical for Lyme Borreliosis." *Clinical Microbiology and Infection* 22, no. 3 (2016): 267.e9–15.

Samuel, B. 2004. *White as a Ghost, Winter Ticks and Moose*. Edmonton: The Federation of Alberta Naturalists.

Sapi, E., K. Balasubramanian, A. Poruri, et al. "Evidence of *In Vivo* Existence of *Borrelia* Biofilm in Borrelial Lymphocytomas." *European Journal of Microbiology & Immunology* 6, no. 1 (2016): 9–24.

Sapi, E., N. Pabbati, A. Datar, et al. "Improved Culture Conditions for the Growth and Detection of *Borrelia* from Human Serum." *International Journal of Medical Sciences* 10, no. 4 (2013): 362–76.

Schulze, T.L., R.A. Jordan, M. Williams, et al. "Evaluation of the SELECT Tick Control System (TCS), a Host-Targeted Bait Box, to Reduce Exposure to *Ixodes Scapularis* (Acari: Ixodidae) in a Lyme Disease Endemic Area of New Jersey." *Journal of Medical Entomology* 54, no. 4 (2017): 1019–24.

Scott, J.D., J.F. Anderson, and L.A. Durden. "Widespread Dispersal of *Borrelia Burgdorferi*-Infected Ticks Collected from Songbirds across Canada." *Journal of Parasitology* 98, no. 1 (2012): 49–59.

Scott, J.D., J.E. Foley, K.L. Clark, et al. "Established Population of Black-legged Ticks with High Infection Prevalence for the Lyme Disease Bacterium, *Borrelia Burgdorferi* Sensu Lato, on Corkscrew Island, Kenora

District, Ontario." *International Journal of Medical Sciences* 13, no. 11 (2016): 881–91.

Sharma, B., A.V. Brown, N.E. Matluck, et al. "*Borrelia Burgdorferi,* the Causative Agent of Lyme Disease, Forms Drug-Tolerant Persister Cells." *Antimicrobial Agents and Chemotherapy* 59, no. 8 (2015): 4616–24.

Shaw, D.K., M. Kotsyfakis, and J.H.F. Pedra. "For Whom the Bell Tolls (and Nods): Spit-Acular Saliva." *Current Tropical Medicine Reports* 3, no. 2 (2016): 40–50.

Sherr, V.T. "Munchausen's Syndrome by Proxy and Lyme Disease: Medical Misogyny or Diagnostic Mystery?" *Medical Hypotheses* 65, no. 3 (2005): 440–47.

Sherrer, Kate. "Target Lesions, Symptoms Not Always Indicative of Lyme Disease." 2016. *Healio News.* November 25.

Smith, R.P., R.T. Schoen, D.W. Rahn, et al. "Clinical Characteristics and Treatment Outcome of Early Lyme disease in Patients with Microbiologically Confirmed Erythema Migrans." *Annals of Internal Medicine* 136, no. 6 (2002): 421–28.

Sonenshine, D.E., and T.N. Mather. *Ecological Dynamics of Tick-Borne Zoonoses.* Oxford, UK: Oxford University Press, 1994.

Sperling, J.L.H., M.J. Middelveen, D. Klein, et al. "Evolving Perspectives on Lyme Borreliosis in Canada." *Open Neurology Journal* 6 (2012): 94–103.

Springer, Y.P., C.S. Jarnevich, D.T. Barnett, et al. "Modeling the Present and Future Geographic Distribution of the Lone Star Tick, *Amblyomma Americanum* (Ixodida: Ixodidae), in the Continental United States." *American Journal of Tropical Medicine and Hygiene* 93, no. 4 (2015): 875–90.

Steere, A.C., G. McHugh, N. Damle, et al. "Prospective Study of Serologic Tests for Lyme Disease." *Clinical Infectious Diseases* 47, no. 2 (2008): 188–95.

Steere, A. C., F. Strle, G. P. Wormser, L. T. Hu, J. A. Branda, J. W. R. Hovius, X. Li, and P. S. Mead. 2016. "Lyme Borreliosis." *Nature Reviews.* Disease Primers 2 (December): nrdp2016090.

Steere, A.C., E. Taylor, G.L. McHugh, et al. "The Overdiagnosis of Lyme Disease." *Journal of the American Medical Association* 269, no. 14 (1993): 1812–26.

Straubinger, R.K., B.A. Summers, Y. Chang, et al. "Persistence of *Borrelia Burgdorferi* in Experimentally Infected Dogs after Antibiotic Treatment." *Journal of Clinical Microbiology* 35, no. 1 (1997): 111–16.

Stricker, R.B., A.F. Corson, and L. Johnson. "Reinfection versus Relapse in Patients with Lyme Disease: Not Enough Evidence." *Clinical Infectious Diseases* 46, no. 6 (2008): 950.

Strle, F., R.B. Nadelman, J. Cimperman, et al. "Comparison of Culture-Confirmed Erythema Migrans Caused by *Borrelia Burgdorferi* Sensu Stricto in New York State and by *Borrelia Afzelii* in Slovenia." *Annals of Internal Medicine* 130, no. 1 (1999): 32–36.

Strle, F., E. Ruži -Sablji , J. Cimperman, et al. "Comparison of Findings for Patients with *Borrelia Garinii* and *Borrelia Afzelii* Isolated from Cerebrospinal Fluid." *Clinical Infectious Diseases* 43, no. 6 (2006): 704–10.

Stromdahl, E., and G. Hickling. "Beyond Lyme: Aetiology of Tick-Borne Human Diseases with Emphasis on the South-Eastern United States." *Zoonoses and Public Health* 59 (2012): 48–64.

Stromdahl, E., R. Nadolny, J. Gibbons, et al. "*Borrelia Burgdorferi* Not Confirmed in Human-Biting *Amblyomma Americanum* Ticks from the Southeastern United States." *Journal of Clinical Microbiology* 53, no. 5 (2015): 1697–704.

Stromdahl, E., M.P. Randolph, J. O'Brien, et al. "*Ehrlichia Chaffeensis* (Rickettsiales: Ehrlichieae) Infection in *Amblyomma Americanum* (Acari: Ixodidae) at Aberdeen Proving Ground, Maryland." *Journal of Medical Entomology* 37, no. 3 (2000): 349–56.

Sumilo, D., L. Asokliene, A. Bormane, et al. "Climate Change Cannot Explain the Upsurge of Tick-borne Encephalitis in the Baltics." *PLoS One* 2, no. 6 (2007): e500.

Summerton, N. "Lyme Disease in the Eighteenth Century." *British Medical Journal* 311, no. 7018 (1995): 1478.

Sykes, R.A., and P. Makiello. "An Estimate of Lyme Borreliosis Incidence in Western Europe." *Journal of Public Health* 39, no. 1 (2017): 74–81.

Tager, F.A., B.A. Fallon, J. Keilp, et al. "A Controlled Study of Cognitive Deficits in Children with Chronic Lyme Disease." *Journal of Neuropsychiatry and Clinical Neurosciences* 13, no. 4 (2001): 500–507.

TäLleklint, L., and T.G.T. Jaenson. "Increasing Geographical Distribution and Density of *Ixodes Ricinus* (Acari: Ixodidae) in Central and Northern Sweden." *Journal of Medical Entomology* 35, no. 4 (1998): 521–6.

Tijsse-Klasen, E., J.J. Jacobs, A. Swart, et al. "Small Risk of Developing Symptomatic Tick-borne Diseases Following a Tick Bite in the Netherlands." *Parasites & Vectors* 4, no. 17 (2011): 1–8.

Tilly, K., P.A. Rosa, and P.E. Stewart. "Biology of Infection with *Borrelia Burgdorferi*." *Infectious Disease Clinics of North America* 22, no. 2 (2008): 217–34.

Tokarz, R., T. Tagliafierro, D.M. Cucura, et al. "Detection of *Anaplasma Phagocytophilum, Babesia Microti, Borrelia Burgdorferi, Borrelia Miya-*

motoi, and Powassan Virus in Ticks by a Multiplex Real-Time Reverse Transcription-PCR Assay." *MSphere* 2, no. 2 (2017): 1–5. e00151-17.

Tull, R., C. Ahn, A. Daniel, et al. "Retrospective Study of Rocky Mountain Spotted Fever in Children." *Pediatric Dermatology* 34, no. 2 (2017): 119–23.

Tutolo, J.W. "Notes from the Field: Powassan Virus Disease in an Infant—Connecticut, 2016." *Morbidity and Mortality Weekly Report* 66, no. 15 (2017): 408-409.

Van Der Weide, L. *"I'm So Grateful That I've Had Her for 39 Years."* (Translated from Dutch). Nouveau, 2015.

van Nunen, S. "Tick-Induced Allergies: Mammalian Meat Allergy, Tick Anaphylaxis and Their Significance." *Asia Pacific Allergy* 5, no. 1 (2015): 3–16.

Van Wieren, S., H. Sprong, W. Takken, et al., eds. *Ecology and Prevention of Lyme Borreliosis*. The Netherlands: Wageningen Academic Publishers, 2016.

Weber, K., and W. Burgdorfer, eds. *Aspects of Lyme Borreliosis*. Berlin, Heidelberg: Springer-Verlag, 1993.

Weintraub, P. *Cure Unknown: Inside the Lyme Epidemic*. New York: St. Martin's Press, 2008.

Weitzner, E., D. McKenna, J. Nowakowski, et al. "Long-term Assessment of Post-Treatment Symptoms in Patients With Culture-Confirmed Early Lyme Disease." *Clinical Infectious Diseases* 61, no. 12 (2015): 1800–1806.

Wikel, S. "Ticks and Tick-Borne Pathogens at the Cutaneous Interface: Host Defenses, Tick Countermeasures, and a Suitable Environment for Pathogen Establishment." *Frontiers in Microbiology* 4, November (2013): 337.

Willyard, C. "Resurrecting the 'Yuppie Vaccine.'" *Nature Medicine* 20, no. 7 (2014): 698–701.

World Health Organization. 2003. *Climate Change and Human Health—Risks and Responses*.

Wormser, G.P., P.J. Baker, S. O'Connell, et al. "Critical Analysis of Treatment Trials of Rhesus Macaques Infected with *Borrelia Burgdorferi* Reveals Important Flaws in Experimental Design." *Vector-Borne and Zoonotic Diseases (Larchmont, N.Y.)* 12, no. 7 (2012): 535–38.

Wormser, G.P., D. Brisson, D. Liveris, et al. "*Borrelia Burgdorferi* Genotype Predicts the Capacity for Hematogenous Dissemination during Early Lyme Disease." *Journal of Infectious Diseases* 198, no. 9 (2008): 1358–64.

Wormser, G.P., R.J. Dattwyler, E.D. Shapiro, et al. "The Clinical Assessment, Treatment, and Prevention of Lyme Disease, Human Granulocytic Anaplasmosis, and Babesiosis: Clinical Practice Guidelines by the Infectious

Diseases Society of America." *Clinical Infectious Diseases* 43, no. 9 (2006): 1089–134.

Wormser, G.P., D. Liveris, K. Hanincová, et al. "Effect of *Borrelia Burgdorferi* Genotype on the Sensitivity of C6 and 2-Tier Testing in North American Patients with Culture-Confirmed Lyme Disease." *Clinical Infectious Diseases* 47, no. 7 (2008): 910–14.

Yoon, E.C., E. Vail, G. Kleinman, et al. "Lyme Disease: A Case Report of a 17-Year-Old Male with Fatal Lyme Carditis." *Cardiovascular Pathology* 24, no. 5 (2015): 317–21.

Yrjänäinen, H., J. Hytönen, P. Hartiala, et al. "Persistence of Borrelial DNA in the Joints of *Borrelia Burgdorferi*-Infected Mice after Ceftriaxone Treatment." *APMIS: Acta Pathologica, Microbiologica, et Immunologica Scandinavica* 118, no. 9 (2010): 665–73.

Zeidner, N.S., R.F. Massung, M.C. Dolan, et al. "A Sustained-Release Formulation of Doxycycline Hyclate (Atridox) Prevents Simultaneous Infection of *Anaplasma Phagocytophilum* and *Borrelia Burgdorferi* Transmitted by Tick Bite." *Journal of Medical Microbiology* 57, no. 4 (2008): 463–8.

Zhou, X., S. Xia, J. Huang, et al. "Human Babesiosis, An Emerging Tickborne Disease in the People's Republic of China." *Parasites & Vectors* 7, no. 509 (2014):1–10.

Index